Adaptation to Climate Change in Agriculture

Toshichika Iizumi • Ryuichi Hirata
Ryo Matsuda
Editors

Adaptation to Climate Change in Agriculture

Research and Practices

Springer

Editors
Toshichika Iizumi
Institute for Agro-Environmental Sciences,
National Agriculture and Food Research
Organization (NARO)
Tsukuba, Ibaraki, Japan

Ryuichi Hirata
Center for Global Environmental Research
National Institute for Environmental Studies
Tsukuba, Ibaraki, Japan

Ryo Matsuda
Graduate School of Agricultural
and Life Sciences
The University of Tokyo
Bunkyo, Tokyo, Japan

ISBN 978-981-13-9234-4 ISBN 978-981-13-9235-1 (eBook)
https://doi.org/10.1007/978-981-13-9235-1

This Springer imprint is published by the registered company Springer Nature Singapore Pte Ltd.
The registered company address is: 152 Beach Road, #21-01/04 Gateway East, Singapore 189721, Singapore

Foreword

The accumulated evidence indicates that climate change has already affected human and natural systems. Climate change mitigation and adaptation are therefore an unavoidable challenge for mankind in the coming decades. For climate change mitigation, international climate negotiations and associated top-down goal setting of greenhouse gas emission reduction are operative. In contrast, for climate change adaptation, local physical and social environments play an important role in forming adaptation measures. Bottom-up approaches based on farmers' innovations and their dissemination should be the key in agricultural adaptation. Researchers, agricultural extension experts, and policymakers are important players in the dissemination of adaptation practices. However, there are distinct knowledge gaps in agricultural adaptation among key players, including policymakers, researchers, extension experts, and farmers. The gaps can be also seen among national and local governments, international organizations, agrobusinesses, and research organizations. An important mission of agricultural meteorology is to provide, improve, and update the scientific knowledge base to bridge these gaps, which enables key players to produce synergies in agricultural adaptation where both top-down approaches led by institutions and bottom-up approaches originated from farmers' innovations need to co-work.

This book is prepared on the occasion of the 75th anniversary of the establishment of The Society of Agricultural Meteorology of Japan, which was started in 1942 to promote the scientific understanding of physical and biological processes in ecosystems, in particular managed ecosystems such as agricultural land. Three board members of the society, Dr. Toshichika Iizumi, Dr. Ryuichi Hirata, and Dr. Ryo Matsuda, proposed the publication plan of this book and served as editors. A vice-chairman, Dr. Yoshiaki Kitaya, provided advisory support for the editors to revise and complete the plan. Financial support in the editing process of this publication was provided by a fund donated by an honorary member, Dr. Taichi Maki. I hope that this book significantly contributes to identifying the role of agricultural meteorology in climate change adaptation and the mission of the society for building more sustainable and resilient societies.

Dr. Masaharu Kitano
Chairman (2017 –2018)
The Society of Agricultural Meteorology of Japan
Faculty of Agriculture
Kyushu University,
Fukuoka, Japan

Preface

The accumulated evidence indicates that agriculture in the world is being affected by climate change. It is likely that farmers and agrobusinesses have already responded to changes in weather patterns, intendedly in some cases and unintendedly in others. The attribution of emerging adaptation is nevertheless not an easy task because not only climate conditions but also surrounding socioeconomic conditions are simultaneously changing.

This book addresses some aspects of this challenge. The detection and understanding of observed change in climate and cultivation practices are key to ultimately leading to technologies, policies, and institutions that enable players in food value chains to adapt to climate change.

The book stems from a series of scientific discussions conducted mainly in the annual meetings of The Society of the Agricultural Meteorology of Japan during the last decade. Particularly, two reports related to adaptation published in the society's journals (*Journal of Agricultural Meteorology* and *Climate in Biosphere*) constitute the direct roots of this book. One is Hirota et al. (2012) which reports the main outcomes of climate change research in Hokkaido, north Japan, presented in the conference held on September 9, 2011. The other is Iizumi et al. (2018) presenting their opinions on ways moving forward in adaptation research in agricultural meteorology based on the discussion made on March 27, 2017. Some authors of these reports or participants in the discussions committedly contributed to this book as authors.

The book is organized to cover practices on the ground and the state-of-the-art research, both aiming to support adaptation of farmers, agrobusinesses, and national food agencies. Part I of the book presents an overview of emerging adaptation responses in agriculture. This helps readers know that adaptation responses that have already emerged are diverse and wide-ranging. Part II describes the forefronts of adaptation research, including modeling, experiments in field and controlled environments, and literature review. Part III elaborates the national adaptation policy and showcases tools offering a means for adaptation for users, from decision support tools for farmers to climate change risk assessment tools for policymakers. Part IV reports a vivid example of how a specific adaptation measure has become popular among farmers in a region, which highlights the role of farmers, agricultural cooperatives, extension organizations, research institutes, and local governments.

There are many other ongoing research and practices for adaptation which are not covered by this book. Though making a catalog of adaptation measures and tools is beyond the scope of our book, the cases presented here enable readers to capture the recent status of adaptation research and practices in agriculture.

We wish to thank the following reviewers for their generous assistance in the editing process of this book:

Mariko Fujisawa	Zenta Nishio
Masayoshi Futami	Masashi Okada
Keiichi Hayashi	Kei Oyoshi
Shoko Hikosaka	Gen Sakurai
Yasushi Ishigooka	Ryoji Sameshima
Yukiyoshi Iwata	Hiroki Sasaki
Kwang-Hyung Kim	Seiji Shimoda
Wonsik Kim	Kiyoshi Takahashi
Yoshiyuki Kinose	Takahiro Takimoto
Tsuguyoshi Kinoshita	Daisuke Yasutake
Atsushi Maruyama	Ryuhei Yoshida

We also thank Dr. Hitoshi Toritani for his comments in the earlier stage of the publication planning.

Tsukuba and Tokyo, Japan
April 2019

Toshichika Iizumi
Ryuichi Hirata
Ryo Matsuda

References

Hirota T, Nakatsuji T, Hamasaki T et al (2012) Report of global warming forum at Hokkaido. Clim Bios 12:B1–11 (in Japanese)

Iizumi T, Masutomi Y, Takimoto T et al (2018) Emerging research topics in agricultural meteorology and assessment of climate change adaptation. J Agric Meteorol 74:54–59

Contents

Part I
Introduction

Chapter 1
Emerging Adaptation to Climate Change in Agriculture

Toshichika Iizumi

Abstract Agriculture is vulnerable to climate change. Adaptation is therefore necessary to moderate harm or exploit the beneficial opportunities of climate change in agriculture. In the 2015 Paris Agreement, the Conference of the Parties of the United Nations Framework Convention on Climate Change (UNFCCC) states, "parties hereby establish the global goal on adaptation." However, limited information on local adaptation practices hinders a review of the overall progress in the adaptation communication. Therefore, we must collect emerging adaptations. This article reviews the available evidence on agricultural adaptation in response to observed or anticipated climate change. A literature review showed both negative and positive impacts of climate change. The emerged adaptations were diverse and ranged from affecting producers to national governments in scope and from incremental to transformational in type. These responses pose challenges in monitoring, modeling, and assessing adaptations in countries at the global scale and require additional research.

1.1 Introduction

Climate change risk assessments for human and natural systems are the basis for adaptation planning and policymaking. The definition of adaptation used in the United Nations Environment Programme (UNEP 2017) is the following: "In human systems, the process of adjustment to actual or expected climate and its effects to moderate harm or exploit beneficial opportunities. In natural systems, the process of adjustment to actual climate and its effects; human intervention may facilitate adjustment to expected climate." The same definition mentioned above is also seen in the Intergovernmental Panel on Climate Change (IPCC) Working Group II

T. Iizumi (✉)
Institute for Agro-Environmental Sciences, National Agriculture and Food Research Organization (NARO), Tsukuba, Ibaraki, Japan
e-mail: iizumit@affrc.go.jp

T. Iizumi et al. (eds.), *Adaptation to Climate Change in Agriculture*,
https://doi.org/10.1007/978-981-13-9235-1_1

(WG2) Fifth Assessment Report (AR5) (IPCC 2014). The definition of adaptation in agriculture would fall within those of the two systems because agriculture is a coupled human–natural system.

The importance of adaptation as well as that of climate mitigation has been well recognized among policymakers (UNFCCC 2012; FAO 2017) and subsequently UNFCCC states that "parties hereby establish the global goal on adaptation" in the Paris Agreement (Article 7; UNFCCC 2015). A global stocktake will occur every 5 years starting in 2023 and will include a review of the overall progress in achieving the goals (UNEP 2017). The latest outcomes from the international negotiation on the global adaptation goal are found in The Katowice Texts (UNFCCC 2018), which state that continuous and enhanced international support shall be provided to developing countries for the implementation of the adaptation communication, the purpose of which is to increase the visibility and profile of adaptation and its balance with mitigation; strengthen adaptation action and support for developing countries; provide input to the global stocktake; and enhance learning and understanding of adaptation needs and actions. However, information required to progress from the adaptation communication is limited, relative to the information available on mitigation efforts. Understanding local adaptations practices and their upscaling is central to ensure progress of the adaptation communication.

The present article reviews the literature that reports emerged adaptation in agriculture to historical or expected climate change. The definition of adaptation used in this review is based on that of UNEP and IPCC. Because of the human–nature coupled constitution of agriculture, both agronomic and non-agronomic adaptations are within the scope of this review. We accepted various sources of information, including scientific journals, farmer's magazines, and news media to collect information because adaptation practices initiated by producers are not necessarily reported as a scientific paper. Most of the collected literature was published after the publication of IPCC WG2 AR5 (IPCC 2014). Japanese literature (categorized as "gray" literature and not preferred to be cited in IPCC reports) is also included in this review. I conclude by discussing knowledge gaps that must be addressed in future research and potential avenues to move forward.

1.2 Emerging Adaptations to Climate Change

Emerging adaptations in agriculture to climate change are diverse (Fig. 1.1). We categorized individual adaptation practices collected for this review into either of three categories: "incremental," "intermediate," or "transformational" for descriptive purposes. Incremental adaptation indicates adjustments of management to keep operating existing systems and it predominantly relies on existing technologies. Switching management practice from conventional one to another which has already operated in another region of the world (warmer regions in particular) is an example for incremental adaptation. In contrast, transformational adaptation requires a change in the fundamental attributes of human and natural systems (IPCC 2014).

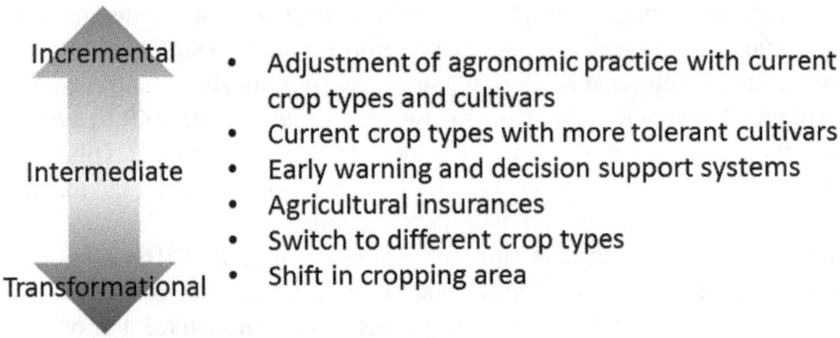

Fig. 1.1 Three qualitative categories of emerging adaptations in agriculture

Conventional management practices would be largely replaced with distinct new ones (e.g., switching crop types) when transformational adaptation occurs (Panda 2018). Moving agricultural areas to other areas is another example for transformational adaptation (Panda 2018). Intermediate adaptation falls within the middle of these two. Importantly, costs for adaptation (including transaction costs, opportunity costs, and others; Rickards and Howden 2012) are expected to be substantial for transformational adaptation compared to those for incremental and intermediate adaptations. Note that the categorization used here is qualitative and currently no quantitative criterion for the separation of the categories is available. These shortfalls are common for earlier studies categorizing adaptation practices (Rickards and Howden 2012; Vermeulen et al. 2013; Panda 2018). The following subsections describe the available evidence of emerging adaptation in agriculture in detail.

1.2.1 Agronomic Adjustments

Agronomic practices have already changed in response to observed climate change. Such agronomic adjustments are deemed as incremental adaptation (Fig. 1.1). Reducing the negative impacts of global warming on the quality of agricultural products (cereals, fruits, vegetables, flowers, forage, and livestock) is an emerging challenge in Japan (Sugiura et al. 2012) as well as in other countries (Nuttall et al. 2017; Ergon et al. 2018). The utilization of heat-tolerant cultivars is a known practice in Japan to moderate the negative impacts of high temperature on rice grain quality and eating quality associated with the occurrence of chalky grain, immature thin grain, and cracked gain (Ishimaru et al. 2016; Morita et al. 2016). In Japan, paddy areas where heat-tolerant rice cultivars are planted have increased from 37,000 hectares (ha) in 2010 (2.3% of the national total paddy area) to 91,400 ha in 2016 (6.6%) (Ministry of Agriculture, Forestry and Fisheries of Japan (MAFF) 2017). Other agronomic practices are recommended to prevent heat damage that affects rice grain quality (Morita et al. 2016), which include water management (irrigation in advance of dry-air-induced water-deficit episodes associated with foehn); fertilizer management (delayed-release fertilizers to moderate heat damage); soil management (deeper tillage and improvements to soil fertility and physical properties which enable rice plants accessing nutrition and water during dry-air-induced water-deficit episodes); and transplanting date shift to avoid risk of poor ripening associated with bad weather (high temperatures, typhoons, etc.). However, we know little about the extent that rice producers adopt the management practices on water, fertilizer, soil, and transplanting.

Similar to transplanting date shift for rice (Morita et al. 2016; Ishigooka et al. 2017), sowing date shift is a known incremental adaptation practice for upland crops (Shimoda et al. 2018). This strategy works well in general, but occasionally fails to moderate the negative impacts of climate change. High summer temperatures are the likely cause of the recent decrease in potato yields in Hokkaido. Shimoda et al. (2018) report that early planting is adopted by producers for adaptation, but it does not necessarily lead to an earlier sprout or elongated growth period. The different climate warming trends (no significant spring warming versus significant summer warming) observed in that region explain in part the difficulty in adaptation solely through sowing date shift. For upland crops, recently, producers in many parts of Japan have used open ditch, laser level, underdrain, and subsoiling methods to increase the drainage performance of their fields (Farmer's Business 2014) in response to observed increases in heavy rainfall events (Fujibe et al. 2006).

1.2.2 Early Warning and Decision Support Tools

Changes in the intensity, frequency, and duration of weather and climate extremes and associated increases in yield variability are anticipated under a warmer climate (Iizumi et al. 2011; Tigchelaar et al. 2018). Yield variability of major crops in 9–22% of the global harvested area has significantly increased during the period 1981–2010. However, decreased yield variability due to technological improvements has been a primary trend for many parts (19–33%) of the world (Iizumi and Ramankutty 2016), with uncertainties in the estimated area shares associated with different yield datasets and spatial resolutions (Iizumi et al. 2018b). Continual monitoring and improvements to preparedness are necessary, as the United Nations Food and Agriculture Organization (FAO) reported that climate variability is a key driver behind the recent rise in global hunger and is one of the leading causes of several food crises in food insecure countries (Porter et al. 2014; FAO 2018).

Reducing the vulnerability and exposure to present-day climate variability is a first step toward adaptation (IPCC 2014). Early responses of food agencies and agrobusinesses to climate extremes are important to moderate climate-induced shock on food supply and commodity market prices, especially in import-dependent countries. Thus, satellite remote sensing, weather predictions, and seasonal climate forecasts are powerful technologies. Food agencies and agrobusinesses receive merits from early warning tools using these technologies to further strengthen their preparedness to climate-induced food supply shocks. Producers also receive merits from weather-related decision support tools based on the technologies to achieve data-armed agriculture, which could be more efficient in fertilizer and chemical uses in unprecedented weather patterns under changing climate than conventional farming. Because technological improvements lead to changes in users' daily operation, a use of early warning and decision support tools is more transformational than the response solely based on conventional technologies (therefore these are deemed as intermediate adaptation) (Fig. 1.1).

Early response recommendations for national food agencies, such as those provided by the FAO when El Niño occurred in 2015/2016 (FAO 2016), are a good practice. Japan has attempted to predict crop production shock in food-exporting countries when El Niño was anticipated to occur in 2014 (see Iizumi et al. (2018c) for more details). Satellite remote sensing, weather predictions, and seasonal climate forecasts have increasingly been used by food agencies as the means for adapting to climate change as well as for moderating negative impacts of climate variability and extremes (Sheffield et al. 2014; Chun et al. 2016; Oyoshi et al. 2016; Vrieling et al. 2016; Iizumi et al. 2018c). Additionally, agro-meteorological information dissemination tools have been designed and operated to support weather-related decision-making for producers and extension services (Ohno et al. 2016; Hayashi et al. 2018; Project SARAI 2018). However, the precision of data at the field scale and the maintenance after the end of the project-oriented system development phase remain a challenge. Research is conducted to enable producers taking a more proactive response to prevent poor skin coloration of grape berries in Japan (Sugiura et al. 2018).

1.2.3 Transformational Adaptations

Some non-agronomic adaptation practices have also been found. For instance, agricultural insurance has been increasingly recognized as an important institution to enable the survival of producers after severe crop failures associated with weather and climate extremes. Both national governments and private companies (e.g., Sompo Japan Nipponkoa Holdings, Inc. (2015)) have started index-based insurance programs in Thailand, Myanmar, Philippines, Indonesia, and Kenya (i.e., insurance payouts for producers are determined by a measured weather index or satellite-derived vegetation index; Greatrex et al. 2015, Vrieling et al. 2016). Because agricultural insurance programs recently introduced are based on new technologies (e.g., index-triggering payouts which are more labor-saving and heavily rely on weather observation networks and/or satellite remote sensing), in this review, these are deemed as transformational adaptation. Different researchers may classify them into different categories.

As agricultural insurance provides financial support to producers, which enables them to shift to new crop types with less risk, they provide safety nets that help increase the producer's adaptive capacity (Fujisawa and Kobayashi 2013) and achieve more transformational adaptation. Switching crop or fruit types is categorized as a transformational adaptation (Fig. 1.1). Fujisawa and Kobayashi (2013) report that the majority of peach producers in Kazuno city (Akita prefecture, northern Japan) switched from apples to peaches once institutional support from the local government and the producer's cooperative, such as subsidies to buy seedlings, establishment of shipping infrastructures, and guidance and training in cultivation methods, became available.

Hypothetically, producers must shift their cropping areas geographically in response to climate change. This is also considered a transformational adaptation (Fig. 1.1). Although current evidence is weak, it appears that cropping areas in western and eastern Australia appear to be shifting south where moisture conditions are better than in inland areas (Hughes 2017). Recent studies reveal that weather and climate extremes play a role in year-to-year variations in cropping areas (Iizumi and Ramankutty 2015; Cohn et al. 2016; Lesk et al. 2016). However, actual changes in cropping area in response to climate change are not well understood.

1.2.4 Positive Influences

Positive impacts of climate change in agriculture have been observed. However, changes in demand, policy, and other socioeconomic conditions accompany and, therefore, in many cases, compound our assessment of climate change adaptation. A clear positive impact includes an elongated growing season. In elevated areas such as the Tibetan Plateau, the warming in spring enables producers to plant the first crop (winter barley) earlier, which allows producers to grow the second crop (rapeseed) during the remaining portion of the growing season to improve soil fertility and income (Zhang et al. 2013). The observed warming enables the harvesting of forage maize twice a year in central Japan (Kanno et al. 2015). The warming also benefits forage rice production in western Japan (Nakano et al. 2009; Sakai et al. 2013) and rice production in central China (Dong et al. 2017) by increasing the annual number of harvests with ratoon-rice cropping.

The production of tropical fruits, for example, mango in Miyazaki prefecture, western Japan, increased from 45.5 ha in 2003 to 93.9 ha in 2015 (MAFF 2015). Climate warming contributes to more favorable conditions for tropical fruit production in that region, although a decrease in the demand for mandarin oranges is the main driver that forced producers to change their fruit type (mandarin orange production in Miyazaki decreased from 977.6 ha in 2003 to 661.7 ha in 2015). Furthermore, reported beneficial opportunities include vine grape production in Hokkaido, northern region (Nagata et al. 2014; Nemoto et al. 2016; Hirota et al. 2017). The record hot summer of 2010 in Hokkaido (2.2 °C higher than the average in 1981–2010) allowed crop scientists to successfully harvest the rice cultivar "Hitomebore," in their experimental field, which is cultivated in warmer regions (Nemoto et al. 2011).

Based on a survey of public agricultural research institutes, Sugiura et al. (2012) reported that 16 out of 47 prefectures in Japan recognize that climate warming has led to a decline in heating costs for vegetable and flower production during the winter; and four to five prefectures recognize increases in wintertime chicken meat and egg production and grass yield. Apple farms in Japan observed earlier blooming times, and higher temperatures during the maturation period likely decrease the acid concentration, fruit firmness, and watercore development, whereas in some cases the soluble-solids concentration increased (Sugiura et al. 2013). These changes are not an overall benefit; however, they positively influence some aspects of the eating quality of apples (Table 1.1).

Table 1.1 A summary of emerging agronomic and non-agronomic adaptations

Adaptation types	Emerging adaptations
Relatively incremental	*Agronomic*
	Heat-tolerant rice cultivars (Ishimaru et al. 2016; Morita et al. 2016; MAFF 2017)
	Adjustments of water, fertilizer, and soil management for paddy rice (Morita et al. 2016)
	Transplanting date shift for paddy rice (Morita et al. 2016; Ishigooka et al. 2017)
	Sowing date shift for potato (Shimoda et al. 2018)
	Open ditch, laser level, underdrain, and subsoiling methods to increase the field's drainage performance (Farmer's Business 2014)
Relatively intermediate	*Non-agronomic*
	Early warning tools for food agencies and agrobusinesses based on satellite remote sensing, weather predictions, and seasonal climate forecasts (Sheffield et al. 2014; Chun et al. 2016; FAO 2016; Oyoshi et al. 2016; Vrieling et al. 2016; Iizumi et al. 2018c)
	Weather-related decision support tools for producers and extension services (Hayashi et al. 2018; Sugiura et al. 2018)
	Agro-meteorological information dissemination tools for producers and extension services (Ohno et al. 2016; Project SARAI 2018)
Relatively transformational	*Agronomic*
	Switching from single cropping to double cropping (rice, Nakano et al. 2009, Sakai et al. 2013, Zhang et al. 2013, Dong et al. 2017; and forage maize, Kanno et al. 2015)
	Switch to tropical fruit types (e.g., mango; MAFF 2015)
	Switch to crop types that are grown in a warmer region (e.g., vine grape; Nagata et al. 2014, Nemoto et al. 2016, Hirota et al. 2017)
	Cropping area shift (Hughes 2017)
	Non-agronomic
	Agricultural and livestock insurances (Greatrex et al. 2015; Sompo Japan Nipponkoa Holdings, Inc. 2015; Vrieling et al. 2016)

1.3 Knowledge Gaps

The following subsections describe important knowledge gaps that need to be addressed in future research. Potential ways of moving forward, if any, are also discussed.

1.3.1 Driving Factors for Adaptation

What are driving factors for adaptation? It is known that producer's perceptions to climate change are important (Fujisawa and Kobayashi 2013, Arbuckle et al. 2015, Bedeke et al. 2018). For instance, the soil-frost depth in Hokkaido has significantly decreased due to increased snow cover in early winter (the adiabatic effect of snow cover prevents deep soil frost) (Hirota et al. 2006), which leads to a serious weed problem associated with volunteer potatoes (Yanai et al. 2017). In that region, some producers started controlling the soil-frost depth by removing snow cover to kill unharvested potato tubers during the winter (snow plowing, or "Yukiwari"). This practice has become common among producers in that region during the last decade (Yazaki et al. 2013; Shimoda et al. 2015; Yanai et al. 2017).

In addition, the producer's sales strategy or their dependence on the farmers' cooperative in selling their products is also important in determining how the producer responds to climate change. Fujisawa et al. (2015) revealed that in Nagano, central Japan apple producers who sold most of their products to the wholesale markets selected accelerating the coloring of apple fruits by placing reflective materials on ground and/or picking off leaves around the fruits to meet the appearance quality standards in the wholesale markets, whereas those who directly sold their products to the consumers selected to simply delay the harvest to full maturity.

1.3.2 Distinctions Between Adaptation and Other Management

A difficulty in addressing questions from climate funding agencies which provide financial supports for climate mitigation and adaptation is the fact that whether a specific agronomic (or non-agronomic) practice is for adaptation or not depends on variables of interest. For instance, an increase in fertilizer input is often misclassified as an adaptation practice when crop yield is a variable of interest in risk assessment. An increase in fertilizer input would lead to the same yield gains between warming and non-warming conditions, and this indicates that yield losses associated with warming is not moderated by this practice (Lobell 2014). In contrast, the decline in rice grain quality in western Japan is attributed to an increase in the average temperature over 20 days after heading, and additional fertilizer input during heading could improve the grain quality (Morita et al. 2016). In this case, an increase in fertilizer input is classified as an adaptation practice. Caution is therefore necessary to make a distinction between adaptation practices and conventional high-yield management for climate funding agencies, of which primary target is mitigation and adaptation but not agricultural development in developing world. However, reported producer's responses in developing regions are often hard to make a distinction (Monirul Alam et al. 2017) because developing countries have many issues need to be solved, and climate change adaptation is one of them. For this issue, efforts by politicians for aligning financial flows with the objectives of the Paris Agreement (mitigation and adaptation) and the United Nations Sustainable Development Goals are ongoing (UNFCCC 2018).

1.3.3 How Can Crop Model Help the Adaptation Communication?

Modeling adaptation is an essential step to improve climate change risk assessments in coming years. However, adaptation practices are diverse in type and level, as previously elaborated. Many risk assessments using eco-physiological process-based crop models consider only a few types of adaptation practices, such as sowing date shift and choice of appropriate cultivar because of the lack of information. However, obviously, this is an unacceptable shortfall to assess progress of adaptation at national and global scales which is a required input for the adaptation communication. Empirical approaches could allow one to estimate the contributions of historical adaptation to crop yields, especially when enormous amounts of data are available (Moore and Lobell 2014). However, this approach provides no explicit information about how and what type and level of adaptation have contributed to produce harmful or beneficial effects of climate change. Other approaches, such as agro-climatic indices (Trnka et al. 2014) that fall within process-based modeling and empirical regressions in terms of model complexity and synthesis based on meta-analysis (Makowski et al. 2015) or multi-method ensembles (Zhao et al. 2017), may be useful to bridge assessment results derived from the two different approaches. Iizumi et al. (2018a) discussed how the insights of adaptation can be derived from crop models with different complexities. To tackle this problem, we require a better understanding of producers' adaptation processes. The concept known as the diffusion of innovations (Rogers 2003) could be useful, as discussed by Fujisawa and Kobayashi (2013).

1.3.4 Adaptation Costs

Difficulties in modeling adaptation hinder the estimation of adaptation costs. Plausible estimates of adaptation costs and residual damages of climate change for a system of interest are central to planning policies, including how to allocate resources among adaptation and mitigation options. Some estimates have been made of global adaptation costs in agriculture under climate change (Wreford and Renwick 2012). However, more detailed information on adaptation costs is desired. For instance, estimated costs of agricultural chemicals should be monitored, given concerns that crop pests and pathogens are shifting poleward (Bebber et al. 2013) and risk of storage insect pests and aflatoxin occurrence in agricultural products will increase under warming (Battilani et al. 2016; Mlambo et al. 2017).

1.3.5 Food Sources Other than Cereal Crops

More importantly, available evidence of adaptations in food production subsectors other than crops is still limited. Cereal crops reviewed in this article are a predominant source of calories, but legumes, fruits, vegetables, livestock products, and fishery products are also important from the nutrition point of view.

1.4 Conclusions

Adaptation in agriculture is occurring worldwide. Although most adaptation practices emerged to date are not necessarily proactive, growing attempts to utilize weather and climate forecast information at food agencies and agrobusinesses are observed. Long-term climate projection information is rarely used for local adaptation in agriculture. Emerging adaptations are diverse in type and level, which poses challenges in monitoring, modeling, and assessing adaptation at national and global scales. A better understanding of adaptation processes in response to observed climate change allows stakeholders (e.g., producers, agricultural cooperatives, agrobusinesses, research institutes, and local and national governments) to be more proactive and have organized adaptation to actual and expected climate change. To achieve this goal, we must review the current adaptation progress and collect all available evidence.

Acknowledgments This review benefited from discussions at the 2018 NARO-FFTC-MARCO Symposium: Climate Smart Agriculture for the Small-Scale Farmers in the Asian and Pacific Region, September 2018 in Tsukuba. T.I. was partly supported by the Grant-in-Aid for Scientific Research (B, 16KT0036 and 18H02317; and C, 17K07984) of Japan Society for the Promotion of Science, the Environment Research and Technology Development Fund (S-14) of the Environmental Restoration and Conservation Agency, and the Joint Research Program of Arid Land Research Center, Tottori University (30F2001).

References

Arbuckle JG Jr, Morton LW, Hobbs J (2015) Understanding farmer perspectives on climate change adaptation and mitigation: the roles of trust in sources of climate information, climate change beliefs, and perceived risk. Environ Behav 47:47205–47234

Battilani P, Toscano P, Van der Fels-Klerx HJ et al (2016) Aflatoxin B1 contamination in maize in Europe increases due to climate change. Sci Rep 6:24328. https://doi.org/10.1038/srep24328

Bebber DP, Ramotowski MAT, Gurr SJ (2013) Crop pests and pathogens move polewards in a warming world. Nat Clim Chang 3:985–988

Bedeke SB, Vanhove W, Wordofa MG et al (2018) Perception of and response to climate change by maize-dependent smallholders. Clim Res 75:261–275

Chun JA, Kim S, Lee W-S et al (2016) Assessment of multimodel ensemble seasonal hindcasts for satellite-based rice yield prediction. J Agric Meteorol 72:107–115

Cohn AS, VanWey LK, Spera SA et al (2016) Cropping frequency and area response to climate variability can exceed yield response. Nat Clim Chang 6:601–604

Dong H, Chen Q, Wang W et al (2017) The growth and yield of a wet-seeded rice-ratoon rice system in central China. Field Crops Res 208:55–59

Ergon Å, Seddaiu G, Korhonen P et al (2018) How can forage production in Nordic and Mediterranean Europe adapt to the challenges and opportunities arising from climate change? Eur J Agron 92:97–106

FAO (2016) 2015–2016 El Niño early action and response for agriculture, food security and nutrition. FAO, Rome. http://www.fao.org/3/a-i6049e.pdf. Accessed 3 July 2018

FAO (2017) Addressing agriculture, forestry and fisheries in national adaptation plans [supplementary guidelines]. FAO, Rome. http://www.fao.org/3/a-i6714e.pdf. Accessed 2 Oct 2018

FAO (2018) The state of food security and nutrition in the world 2018. Building climate resilience for food security and nutrition. FAO, Rome. http://www.fao.org/3/I9553EN/i9553en.pdf. Accessed 2 Oct 2018

Farmer's Business (2014) Farming to prepare abnormal weather: risk management under the time of abnormal weather. 218:17–29. https://agri-biz.jp/item/detail/7961. Accessed 2 Oct 2018 (in Japanese)

Fujibe F, Yamazaki N, Kobayashi K (2006) Long-term changes of heavy precipitation and dry weather in Japan (1901–2004). J Meteorol Soc Japan 84:1033–1046

Fujisawa M, Kobayashi K (2013) Shifting from apple to peach farming in Kazuno, Northern Japan: perceptions of and responses to climatic and non-climatic impacts. Reg Environ Chang 13:1211–1222

Fujisawa M, Kobayashi K, Johnston P et al (2015) What drives farmers to make top-down or bottom-up adaptation to climate change and fluctuations? A comparative study on 3 cases of apple farming in Japan and South Africa. PLoS One 10(3):e0120563. https://doi.org/10.1371/journal.pone.0120563

Greatrex H, Hansen JW, Garvin S et al (2015) Scaling up index insurance for smallholder farmers: recent evidence and insights. CGIAR research program on climate change, agriculture and food security (CCAFS), Copenhagen. https://ccafs.cgiar.org/publications/scaling-index-insurance-smallholder-farmers-recent-evidence-and-insights. Accessed 25 Sept 2018

Hayashi K, Llorca L, Rustini S et al (2018) Reducing vulnerability of rainfed agriculture through seasonal climate predictions: a case study on the rainfed rice production in Southeast Asia. Agric Syst 162:66–76

Hirota T, Iwata Y, Hayashi M et al (2006) Decreasing soil-frost depth and its relation to climate change in Tokachi, Hokkaido, Japan. J Meteorol Soc Japan 84:821–833

Hirota T, Yamazaki T, Yasui Y et al (2017) Climate change favors viticulture and wine production in Hokkaido, Japan. Clim Biosph 17:34–45

Hughes N (2017) Australian farmers are adapting to climate change. The Conversation Media Group Ltd., Parkville. http://theconversation.com/australian-farmers-are-adapting-to-climate-change-76939. Accessed 11 Jan 2019

Iizumi T, Ramankutty N (2015) How do weather and climate influence cropping area and intensity? Glob Food Sec 4:46–50
Iizumi T, Ramankutty N (2016) Changes in yield variability of major crops for 1981–2010 explained by climate change. Environ Res Lett 11:034003. https://doi.org/10.1088/1748-9326/11/3/034003
Iizumi T, Yokozawa M, Nishimori M (2011) Probabilistic evaluation of climate change impacts on paddy rice productivity in Japan. Clim Chang 107:391–415
Iizumi T, Masutomi Y, Takimoto T et al (2018a) Emerging research topics in agricultural meteorology and assessment of climate change adaptation. J Agric Meteorol 74:54–59
Iizumi T, Kotoku M, Kim W et al (2018b) Uncertainties of potentials and recent changes in global yields of major crops resulting from census- and satellite-based yield datasets at multiple resolutions. PLoS One 13(9):e0203809. https://doi.org/10.1371/journal.pone.0203809
Iizumi T, Shin Y, Kim W et al (2018c) Global crop yield forecasting using seasonal climate information from a multi-model ensemble. CLISER 11:13–23. https://doi.org/10.1016/j.cliser.2018.06.003
IPCC (2014) Summary for policymakers. In: Field CB, Barros VR, Dokken DJ et al (eds) Climate change 2014: impacts, adaptation, and vulnerability. Part A: global and sectoral aspects. IPCC, Cambridge, pp 1–32
Ishigooka Y, Fukui S, Hasegawa T et al (2017) Large-scale evaluation of the effects of adaptation to climate change by shifting transplanting date on rice production and quality in Japan. J Agric Meteorol 73:156–173
Ishimaru T, Hirabayashi H, Sasaki K et al (2016) Breeding efforts to mitigate damage by heat stress to spikelet sterility and grain quality. Plant Prod Sci 19:12–21
Kanno T, Morita S, Sasaki H et al (2015) Recent situation and future prediction for areas suitable for double cropping of corn (*Zea mays L.*) production in Japan's Kanto region. Jpn J Grassl Sci 60:161–166
Lesk C, Rowhani P, Ramankutty N (2016) Influence of extreme weather disasters on global crop production. Nature 529:84–87
Lobell DB (2014) Climate change adaptation in crop production: beware of illusions. Glob Food Sec 3:72–76
MAFF (2015) Local fruit tree production survey. http://www.maff.go.jp/j/tokei/kouhyou/tokusan_kazyu/. Accessed 25 Sept 2018 (in Japanese)
MAFF (2017) Global warming impacts survey report. MAFF, Tokyo. http://www.maff.go.jp/j/seisan/kankyo/ondanka/attach/pdf/index-17.pdf. Accessed 3 July 2018 (in Japanese)
Makowski D, Asseng S, Ewert F et al (2015) A statistical analysis of three ensembles of crop model responses to temperature and CO_2 concentration. Agric For Meteorol 214:483–493
Mlambo S, Mvumi BM, Stathers T et al (2017) Field efficacy of hermetic and other maize grain storage options under smallholder farmer management. Crop Prot 98:198–210
Monirul Alam GM, Alam K, Mushtaq S (2017) Climate change perceptions and local adaptation strategies of hazard-prone rural households in Bangladesh. Clim Risk Manag 17:52–63
Moore FC, Lobell DB (2014) Adaptation potential of European agriculture in response to climate change. Nat Clim Chang 4:610–614
Morita S, Wada H, Matsue Y (2016) Countermeasures for heat damage in rice grain quality under climate change. Plant Prod Sci 19:1–11
Nagata O, Kobahashi K, Niwa K et al (2014) Report for the organized session of the annual meeting of the Society of Agricultural Meteorology of Japan 2014—future perspective regarding the Sorachi area in Hokkaido as a place of wine production. Clim Biosph 14:D18–D28 (in Japanese)
Nakano H, Morita S, Kitagawa H et al (2009) Effects of cutting height and trampling over stubbles of the first crop on dry matter yield in twice harvesting of forage rice. Plant Prod Sci 12:124–127
Nemoto M, Hamasaki T, Sshimono H (2011) Extraordinary hot summer in Hokkaido decrease rice yield and satisfy growing of cultivar in Tohoku region "Hitomebore". J Agric Meteorol 67:269–274

Nemoto M, Hirota T, Sato T (2016) Prediction of climatic suitability for wine grape production under the climatic change in Hokkaido. J Agric Meteorol 72:167–172

Nuttall JG, O'Leary GJ, Panozzo JF et al (2017) Models of grain quality in wheat–a review. Field Crops Res 202:136–145

Ohno H, Sasaki K, Ohara G et al (2016) Development of grid square air temperature and precipitation data compiled from observed, forecasted, and climatic normal data. Clim Biosph 16:71–79

Oyoshi K, Tomiyama N, Okumura T et al (2016) Mapping rice-planted areas using time-series synthetic aperture radar data for the Asia-RiCE activity. Paddy Water Environ 14:463–472

Panda A (2018) Transformational adaptation of agricultural systems to climate change. WIREs Clim Change 9:e520. https://doi.org/10.1002/wcc.520

Porter JR, Xie L, Challinor AJ et al (2014) Food security and food production systems. In: Field CB, Barros VR, Dokken DJ et al (eds) Climate change 2014: impacts, adaptation, and vulnerability. Part A: global and sectoral aspects. IPCC, Cambridge, pp 485–533

Project SARAI (2018) Smart approaches to reinvigorate agriculture as an industry in the Philippines. SARAI training toolkit for rice. The University of the Philippines Los Baños, Los Baños. http://202.92.144.44/images/Eskwela/Toolkit/Introduction%20to%20SARAI%20Training%20Toolkit.pdf. Accessed 2 Oct 2018

Rickards L, Howden SM (2012) Transformational adaptation: agriculture and climate change. Crop Pasture Sci 63:240–250

Rogers EM (2003) Diffusion of innovations, 5th edn. Free Press, New York

Sakai M, Nakano H, Okamoto M et al (2013) "Ruriaoba", a new rice forage variety for double harvesting. Bull NARO Kyushu Okinawa Agric Res 60:1–12

Sheffield J, Wood EF, Chaney N et al (2014) A drought monitoring and forecasting system for sub-Sahara African water resources and food security. Bull Am Meteorol Soc 95:861–882

Shimoda S, Yazaki T, Nishio Z et al (2015) Possible soil frost control by snow compaction on winter wheat fields. J Agric Meteorol 71:276–281

Shimoda S, Kanno H, Hirota T (2018) Time series analysis of temperature and rainfall-based weather aggregation reveals significant correlations between climate turning points and potato (*Solanum tuberosum L*) yield trends in Japan. Agric For Meteorol 263:147–155

Sompo Japan Nipponkoa Holdings, Inc (2015) SOMPO HOLDINGS joins business call to action (BCtA) as the first P&C insurance group in the world. Sompo Japan Nipponkoa Holdings, Inc., Tokyo. http://www.sompo-hd.com/~/media/hd/en/files/news/2015/e_20150731_1.pdf. Accessed 25 Sept 2018

Sugiura T, Sumida H, Yokoyama S et al (2012) Overview of recent effects of global warming on agricultural production in Japan. Jpn Agric Res Q 46:7–13

Sugiura T, Ogawa H, Fukuda N et al (2013) Changes in the taste and textural attributes of apples in response to climate change. Sci Rep 3:2418. https://doi.org/10.1038/srep02418

Sugiura T, Shiraishi M, Konno S et al (2018) Prediction of skin coloration of grape berries from air temperature. Hortic J 87:18–25

Tigchelaar M, Battisti DS, Naylor RL et al (2018) Future warming increases probability of globally synchronized maize production shocks. Proc Natl Acad Sci U S A 115:6644–6649

Trnka M, Rötter RP, Ruiz-Ramos M et al (2014) Adverse weather conditions for European wheat production will become more frequent with climate change. Nat Clim Chang 4:637–643

UNEP (2017) The adaptation gap report 2017. UNEP, Nairobi. https://www.unenvironment.org/resources/report/adaptation-gap-report-2017. Accessed 25 Sept 2018

UNFCCC (2012) National adaptation plans. Technical guidelines for the national adaptation plan process. UNFCCC, Bonn. http://unfccc.int/files/adaptation/cancun_adaptation_framework/application/pdf/naptechguidelines_eng_high__res.pdf. Accessed 2 Oct 2018

UNFCCC (2015) Paris agreement. UNFCCC, Bonn. https://unfccc.int/sites/default/files/english_paris_agreement.pdf. Accessed 25 Sept 2018

UNFCCC (2018) The Katowice texts. Proposal by the president. UNFCCC, Bonn. https://unfccc.int/sites/default/files/resource/Katowice%20text%2C%2014%20Dec2018_1015AM.pdf. Accessed 11 Jan 2019

Vermeulen SJ, Challinor AJ, Thornton PK et al (2013) Addressing uncertainty in adaptation planning for agriculture. Proc Natl Acad Sci U S A 110:8357–8362

Vrieling A, Meroni M, Mude AG et al (2016) Early assessment of seasonal forage availability for mitigating the impact of drought on East African pastoralists. Remote Sens Environ 174:44–55

Wreford A, Renwick A (2012) Estimating the costs of climate change adaptation in the agricultural sector. CAB Rev 7:1–10

Yanai Y, Iwata Y, Hirota T (2017) Optimum soil frost depth to alleviate climate change effects in cold region agriculture. Sci Rep 7:44860. https://doi.org/10.1038/srep44860

Yazaki T, Hirota T, Iwata Y et al (2013) Effective killing of volunteer potato (*Solanum tuberosum L.*) tubers by soil frost control using agrometeorological information—an adaptive countermeasure to climate change in a cold region. Agric For Meteorol 182–183:91–100

Zhang G, Dong J, Zhou C et al (2013) Increasing cropping intensity in response to climate warming in Tibetan Plateau, China. Field Crops Res 142:36–46

Zhao C, Liu B, Piao S et al (2017) Temperature increase reduces global yields of major crops in four independent estimates. Proc Natl Acad Sci U S A 114:9326–9331

Part II
Research Towards Adaptation

Chapter 2
Impact Assessment and Adaptation Simulation for Chalky Rice Grain in the Cultivar 'Koshihikari' in Japan Using Large Ensemble Climate Projection Data Sets

Takahiro Takimoto, Yuji Masutomi, and Makoto Tamura

Abstract In this study, we used a statistical model to simulate the impact of climate change and adaptations on the occurrence of chalky rice grains (CRG), recognized as an indicator of rice quality. The target variety was 'Koshihikari', the most widely grown cultivar in Japan. A large number of ensemble climate projection data sets were used to provide probability information for impact assessment and adaptation simulations. The impact assessment predicted that the occurrence of CRGs would increase with increasing temperature in all regions of Japan. One adaptation simulation indicated that moving the transplanting date would be effective for reducing the occurrence of CRG. A further adaptation simulation using a virtual heat-tolerant variety indicated that this would also be effective in reducing CRG throughout Japan. The results of this study will be useful to stakeholders such as farmers and policymakers when considering the impact of climate change on rice quality.

T. Takimoto (✉)
Institute for Agro-Environmental Sciences, National Agriculture and Food Research Organization, Tsukuba, Ibaraki, Japan
e-mail: t.takimoto@affrc.go.jp

Y. Masutomi
College of Agriculture, Ibaraki University, Ami, Ibaraki, Japan

Institute for Global Change Adaptation Science, Ibaraki University, Mito, Ibaraki, Japan

M. Tamura
Institute for Global Change Adaptation Science, Ibaraki University, Mito, Ibaraki, Japan

T. Iizumi et al. (eds.), *Adaptation to Climate Change in Agriculture*,
https://doi.org/10.1007/978-981-13-9235-1_2

2.1 Introduction

Rice is a major crop in monsoon Asia, and is also an important crop in Japan, where it is grown on 54% of the cultivated area (Ministry of Agriculture, Forestry, and Fisheries, MAFF 2018). Recent climate change has affected rice crops in Japan. In particular, the impact on rice quality has become apparent. Although there are various requirements for defining the quality of rice, the occurrence of chalky rice grains (CRGs) is considered to be the most significant problem.

CRGs are characterized by an opaque region within the endosperm, and represent a problem at each stage from harvest to consumption (Takimoto et al. accepted). In Japan, the percentage of CRGs in a given number of grains is one of the criteria used for determining rice grade, in accordance with Japan's Agricultural Products Inspection Act. Because the accumulated density of starch in CRGs is smaller than that of the undamaged grain, the CRGs tend to be damaged by rice milling (Wang et al. 2007; Morita 2008). In addition, CRGs are not as popular in the market because their texture is less sticky (Yamakawa et al. 2007), and their taste is considered inferior (Cheng et al. 2005; Wakamatsu et al. 2007; Morita 2008). Accordingly, both farmers and governments have a particular interest in the occurrence of CRGs.

Experiments have clarified that the meteorological conditions that promote the development of CRGs in rice are high temperature and low solar radiation during the ripening period (Morita 2005; Kondo 2010). Indeed, conditions during the hot summer of 1999 in Japan resulted in a large number of CRGs (Terashima et al. 2001). Against this background, the challenge we must address at the research level is to assess the impact of CRGs and to develop adaptative measures. Specifically, the following two questions must be answered: (1) To what extent will the occurrence of CRGs change in response to projected climate change? (2) How can we reduce the occurrence of CRGs?

Research on the impact of climate change on rice production in Japan has traditionally tended to focus on yield (e.g. Seino 1997; Iizumi et al. 2011), and in recent years, on the quality of rice. For example, Okada et al. (2011) conducted an impact assessment on rice yield and quality in Kyushu, an island in southern Japan. They targeted the rice grade for quality and used eight or nine climate projection data sets under two emissions scenarios. Similarly, Ishigooka et al. (2017) conducted an impact assessment, including adaptative measures, of rice yield and quality across Japan. They used six general circulation models (GCMs) for each of three emission scenarios as climate projection data, and a risk index proposed by Ishigooka et al. (2011) for rice quality. In these studies, however, the occurrence of CRGs was not directly evaluated.

Including these studies, previous research has used climate projection data to assess impacts and adaptation. The number of climate projection data sets used in these studies was only a few dozen (e.g. Masutomi et al. 2009). However, such ensemble data cannot be used to determine the probability of an impact. If probability information can be incorporated into the impact assessment, stakeholders will be able to assess the impact by taking year-to-year climate variability into account. In recent years, impact assessments in various sectors have been conducted using thousands of ensemble climate projection data sets (e.g. Imada et al. 2017; Yoshida et al. 2017).

The purpose of this study was to evaluate the impact of climate change on the occurrence of CRGs and to simulate adaptative measures in Japan using ensemble climate projection data. The target rice variety was 'Koshihikari', which is the most widely cultivated variety in Japan (Kobayashi et al. 2018). We used a model developed by Takimoto et al. (accepted) to estimate the occurrence of CRGs in conjunction with a large number of ensemble climate projection data sets to evaluate the probability information regarding the impact assessment. We also conducted an adaptation simulation of the reduction in CRGs based on a shift in the transplanting date, as examined by Okada et al. (2011) and Ishigooka et al. (2017). In addition, the adaptative effect of a virtually created heat-tolerant variety was examined. Although it is also possible to reduce CRGs based on cultivation management, such as fertilizer and water management (Morita et al. 2016), we did not examine the effects of these in this study, as our main focus was on meteorological conditions.

2.2 Data and Methodology

2.2.1 *Chalky Rice Grain Model*

We applied a statistical model for estimating the occurrence of CRGs (Takimoto et al. accepted). Details of model development, parameter estimation, and model accuracy are available from Takimoto et al. (accepted). The model is based on the following empirical knowledge of the relationship between CRGs and meteorological factors: CRGs occur when rice is exposed to certain air temperatures; the occurrence of CRGs increases with increasing air temperature; and the occurrence of CRGs changes in response to the amount of solar radiation (Morita 2005; Kondo et al. 2006; Kondo, 2010; Masutomi et al. 2015). On the basis of these assumptions, Takimoto et al. (accepted) proposed the following relationships between CRGs, air temperature, and solar radiation:

$$I(T,S) = \begin{cases} 0 & (T < aS + b) \\ k(T - aS - b) & (T \geq aS + b) \end{cases} \tag{2.1}$$

Here, I (%) is the percentage of CRGs in the harvested grains, T (°C) and S (MJ m^{-2} d^{-1}) are the average daily mean air temperature and average daily accumulation of solar radiation over a given period, respectively, and k (% °C^{-1}) is the sensitivity of the occurrence of CRGs to T. $aS + b$ indicates the air temperature threshold for the occurrence of CRGs as a function of solar radiation. Parameters k, a, and b in 'Koshihikari' were determined based on multi-year experimental results of 6.59 (% °C^{-1}), 0.45 (°C (MJ m^{-2} d^{-1})$^{-1}$), and 16.41 (°C), respectively. The root-mean-square error of the model is 2.55%, and the average period for T and S is 34 days from 5 days before the heading period, respectively (Takimoto et al. accepted). Therefore, by using Eq. 2.1, it is possible to estimate the occurrence of CRGs based on the date of the heading period, daily mean air temperature, and daily accumulation of solar radiation.

2.2.2 Meteorological Data

In this section, we describe the meteorological data used in the present study. Large ensemble climate projection data sets were used for the impact assessment and adaptation simulations. Details of these data sets are provided in Sect. 2.2.2.1. Observation-based gridded meteorological data were used to correct the bias contained in the climate projection data. The details of these data sets and the bias correction method are presented in Sects. 2.2.2.2 and 2.2.2.3, respectively.

2.2.2.1 The Database for Policy Decision Making for Future Climate Change (d4PDF)

For the purposes of this study, we used the database for policy decision making for future climate change (d4PDF database). The d4PDF database contains data related to large historical and future ensemble climate simulations. The global simulations were conducted based on 60 km horizontal grids, using MRI-AGCM, which is a GCM developed by the Meteorological Research Institute (MRI). The most notable feature of the d4PDF database is that it contains a large number of historical and future sea surface temperature (SST) patterns and climate simulations performed as boundary conditions. For future ensemble climate projections, SSTs were generated based on global mean surface air temperatures of 2 and 4 °C higher than pre-industrial conditions (1861–1880). Perturbed SSTs were also prepared for the historical simulations. On the basis of the historical and future ensemble global simulations, a regional climate model was conducted using MRI-NHRCM, which was developed by MRI with 20 km horizontal grids. We used 3240- and 5400-year ensemble data sets for the 2 and 4 °C experiments (hereafter referred to as +2K_TEMP and +4K_TEMP, respectively). In the historical simulations, we used 50 ensembles each year from 1980 to 2010 (hereafter referred to as HR). Further details of the d4PDF database are presented by Mizuta et al. (2017).

Data for daily mean air temperatures and the daily accumulation of solar radiation are required to estimate the occurrence of CRGs. In addition, daily maximum and minimum temperature data are needed to estimate the heading period, as described in Sect. 2.2.3. These meteorological variables were extracted from the d4PDF database. The model simulation output included hourly calculated values, including those for these meteorological variables. The 24-hourly air temperature values obtained for each day were averaged to provide the daily mean temperature. Maximum and minimum values of the 24-hourly air temperature values were defined as maximum and minimum temperature, respectively. The daily accumulation of solar radiation was calculated by integrating solar radiation throughout the day.

2.2.2.2 Agro-Meteorological Grid Square Data

Given that the HR, +2K_TEMP, and +4K_TEMP data contain bias, it was initially necessary to remove this bias from each meteorological variable prior to conducting the impact assessment and adaptation simulations. In this study, we performed bias correction using the Agro-Meteorological Grid Square Data (AMGS; Ohno et al. 2016). AMGS is a database in which ground meteorological observations are spatially interpolated. The Automated Meteorological Data Acquisition System (AMeDAS), which is an observation network of the Japan Meteorological Agency, was used in generating AMGS. AMeDAS carries out meteorological observations in Japan for meteorological factors such as air temperature and sunshine duration, at approximately 1300 sites, which corresponds to one site per 17 km^2. AMGS spatially interpolates these observation values, to produce data on grids with horizontal spacing of approximately 1 km.

Although AMeDAS does not directly measure solar radiation, AMGS estimates solar radiation using a conversion formula based on observed sunshine duration, which are then interpolated across the area. Daily mean, maximum, and minimum temperatures are spatially interpolated using the observed values. In the present study, AMGS were averaged to match the spatial resolution of the d4PDF database. The data were then used for bias correction. The data used were collected during the 31-year period from 1980 to 2010, which corresponds to the years of the HR experiment.

2.2.2.3 Bias Correction

Bias correction was performed for daily mean, maximum, and minimum temperatures, and daily accumulation of solar radiation using the following procedure. First, monthly mean values were calculated using data from the AMGS database and HR experiments, and the difference between the corresponding values was defined as a bias. This bias was subsequently added to the daily HR, +2K_TEMP, and +4K_TEMP values. These corrections conserved the variability of the meteorological variables in the d4PDF database.

2.2.3 Heading Period Model

It is well known that for rice, the period from transplanting to heading depends on temperature and day length. In order to estimate I (%CRGs in the harvested grains), the daily mean air temperature and daily accumulation of solar radiation must be calculated based on the heading period. Therefore, it is necessary to consider changing the heading period along with the changes in air temperatures when conducting impact assessment and adaptation simulations. Some previous studies have developed a model to estimate the heading period (Zhang and Tao 2013). In the present study, we used a model developed by Fukui et al. (2015) to estimate the heading period. Fukui et al. (2015) determined model parameters in a dozen Japanese rice cultivars, including 'Koshihikari'. In addition, verification of the model was conducted nationwide. The daily maximum and minimum air temperatures were required in order to calculate the heading period in the model.

2.2.4 Area Calculations

Although 'Koshihikari' is widely cultivated in Japan, it is not cultivated in all prefectures. Thus, in the present study, we excluded from the analyses those prefectures in which this cultivar is not grown. The MAFF publishes the percentage area of planted varieties in each prefecture (MAFF 2009), and with reference to these data, we excluded the following nine prefectures where the percentage cultivated area of 'Koshihikari' was less than 5%: Hokkaido, Aomori, Iwate, Miyagi, Akita, Osaka, Fukuoka, Oita, and Okinawa. The remaining 38 prefectures were used for the analysis. In addition, we also excluded squares in the d4PDF database grid encompassing small areas of paddy field. For this purpose, we used the 'Land Use Mesh 2006' data set from the National Land Numerical Information (Geographical Survey Institute, Ministry of Land, Infrastructure, Transport, and Tourism of Japan 2006). This data set classifies land use into 11 types, including paddy fields, and provides the percentages of each land use in each cell. We calculated the percentage area of paddy fields in the grids of the d4PDF database, and excluded those grid squares with a percentage area of less than 0.5% (this approximated to less than 2 km^2 per 400 km^2). Such grids mainly covered areas of cities, mountains, and oceans.

Spatial data related to rice transplanting dates were necessary for estimations of the heading periods. Data on the spatial distribution of rice transplanting dates from 1980 to 2010 were obtained from a MAFF data set that provides yearly statistics for cultivation schedules, including the transplanting of rice (Ishigooka et al. 2011, 2017).

In this study, we combined the data for several prefectures to illustrate the results of the impact assessment and adaptation simulations. On the basis of previous studies, the prefectures were aggregated into nine regions as shown in Fig. 2.1 (Ishigooka et al. 2011, 2017).

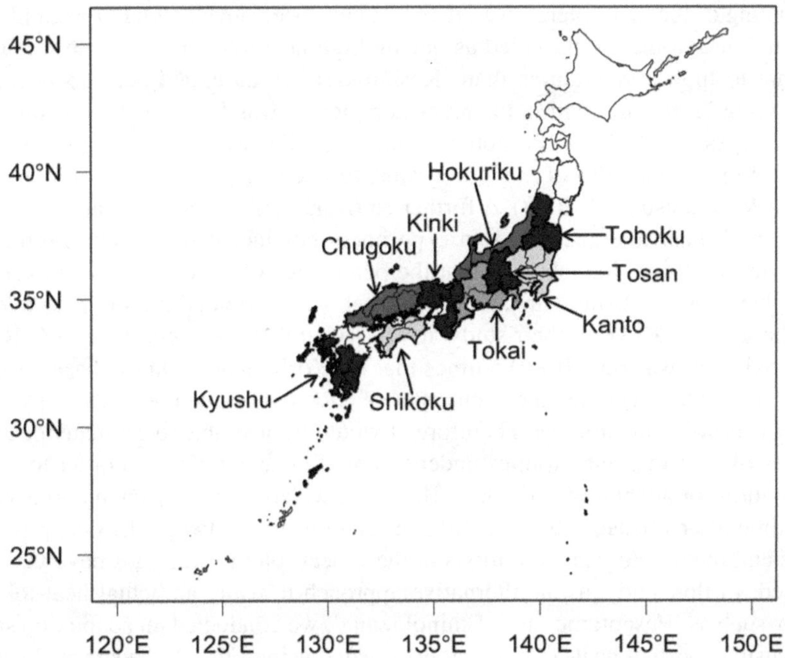

Fig. 2.1 Study regions across Japan. The prefectures shown in white were excluded from the analysis

2.2.5 Calculation Conditions of the Impact Assessment and Adaptation Simulations

An impact assessment was performed for the HR, +2K_TEMP, and +4K_TEMP experiments. In order to calculate the heading period for HR, we used the spatial distribution of rice transplanting dates corresponding to the year. For +2K_TEMP and +4K_TEMP, the transplanting date was fixed as the value in 2010, and then the heading period was calculated from all the ensemble data. On the basis of the projections of +2K_TEMP and +4K_TEMP, it is predicted that the areas and locations of paddy fields in Japan may change in the future. However, as this information is currently unavailable, we assumed that the areas and locations of paddy fields will remain unchanged in the future. Therefore, the calculated grid is the same as that used for HR.

An adaptation simulation with a shifting transplanting date was performed using +2K_TEMP. As with the +2K_TEMP impact assessment, the transplanting date was fixed as the date in 2010. The transplanting date was then adjusted at 7-day intervals to 35 days before and 35 days after the 2010 date, and the heading period was calculated accordingly. Thereafter, we calculated *I*.

According to a recent report of the MAFF, the tolerance of 'Koshihikari' to high temperatures is ranked as intermediate (MAFF 2018). The same report also describes the ranking of the heat-tolerant varieties that have been developed. For example, the cultivar 'Fusaotome' is classified as having high heat tolerance because it is more resistant to high temperatures than 'Koshihikari'. In an experiment conducted in Kagoshima Prefecture, within the Kyushu region, it was found that the white-back grains (types of CRG) of 'Fusaotome' and 'Koshihikari' accounted for less than 30% and more than 70% of the total grains, respectively, at a protein content level of 6% (Wakamatsu et al. 2008). A further cultivar, 'Eminokizuna', is also ranked as having high heat tolerance, and an experiment conducted in Saitama Prefecture, within the Kanto region, showed that the percentage of CRGs in this cultivar was lower than that in 'Koshihikari', indicating that 'Eminokizuna' is more heat-tolerant (Sasahara et al. 2018). In their study, it was found that the occurrence of CRGs in 'Eminokizuna' was only 0.1–0.5 times that in 'Koshihikari'. On the basis of these results, it is anticipated that the occurrence of CRGs in rice can be reduced by introducing heat-tolerant varieties. Therefore, it would be desirable to estimate the effectiveness of heat-tolerant varieties under future climate condition in order to obtain information for adaptative measures. However, we are currently unable to perform such simulations because the quantitative relationships between the occurrence of CRGs and meteorological conditions in these heat-tolerant varieties have yet to be clarified. In this study, as an alternative approach to using an actual heat-tolerant variety such as 'Fusaotome' and 'Eminokizuna', we conducted an adaptation simulation using a virtual heat-tolerant variety created using Eq. 2.1. This virtual variety was created by adding 0.2 °C increments from 1.0 to 2.2 °C (i.e. 1.0, 1.2, 1.4, ..., or 2.2 °C) to the parameter b of 'Koshihikari' in Eq. 2.1. We consider that this range of heat tolerance is realistic because in the aforementioned example of 'Eminokizuna', it was estimated that by adding approximately 1.1 to 3.6 °C to the b value of 'Koshihikari' in Eq. 2.1, we could produce the same occurrence of CRGs as in 'Eminokizuna'. The simulation using the virtual heat-tolerant variety was carried out under the +2K_TEMP meteorological conditions, with the transplanting date set to that in 2010, as in the +2K_TEMP impact assessment. Therefore, the heading period in this simulation was the same as that under +2K_TEMP in the impact assessment. For both adaptation simulations, the grid to be calculated was the same as that in the impact assessment.

2.3 Results and Discussion

2.3.1 Impact Assessment: The Impact of Climate Change on Chalky Rice Grains

Figures 2.2 and 2.3 show the relationships between the percentage of chalky rice grains (I) and the historical mean T over each grid and the temperature differences from the historical mean T (hereafter T_{diff}) in each ensemble. These graphs enable us to assess future changes in T and I from the historical T values with probability

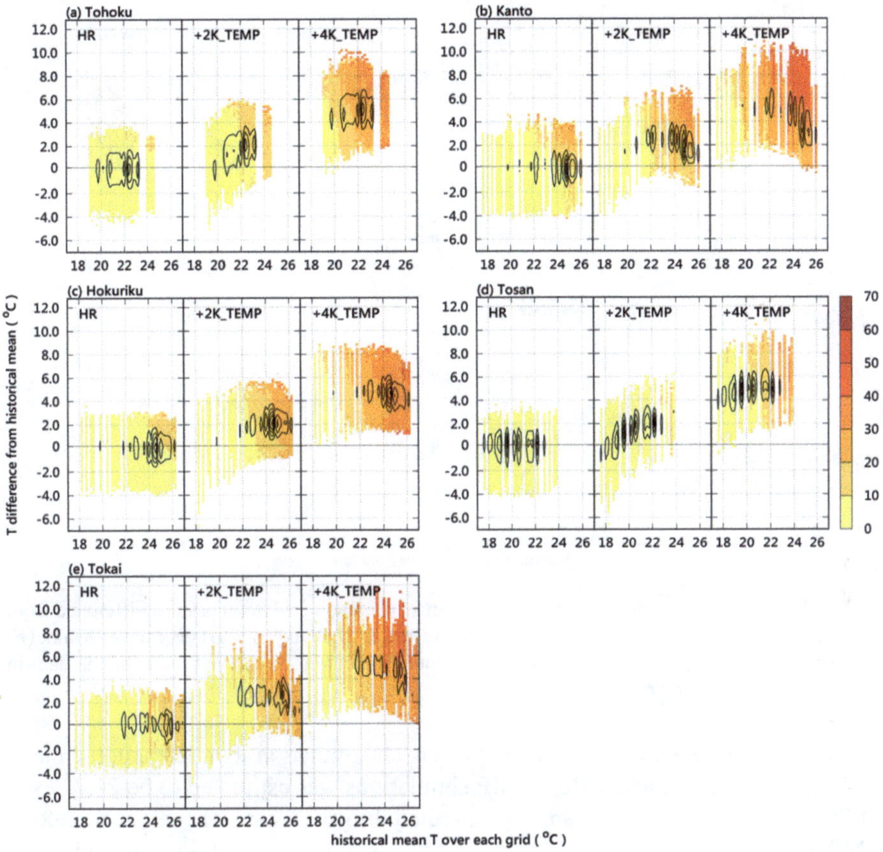

Fig. 2.2 The relationships between historical mean *T* over each grid, temperature difference from the historical mean *T* in each ensemble, and the percentage of chalky rice grains *I* in (**a**) Tohoku, (**b**) Kanto, (**c**) Hokuriku, (**d**) Tosan, and (**e**) Tokai. The contours indicate the relative counts of the data in 0.2% increments from 0.2%

information. For example, when the historical mean *T* was 22 °C in Kanto, T_{diff} and *I* values ranged from approximately −1 to 6 °C and 0% to 20%, respectively. Considering the number of data points in this range, the probable variations in T_{diff} and *I* were 1–4 °C and 10%–20%, respectively. In the +4K_TEMP simulation, T_{diff} values ranging from approximately 5 to 7 °C and *I* values ranging from 20% to 40% were likely to occur. In case of Shikoku, when the historical mean *T* was 22 °C, the T_{diff} and *I* values for +2K_TEMP generally varied from 2 to 4 °C and from 0% to 20%, respectively. In the +4K_TEMP simulation, T_{diff} values ranging from 4 to 6 °C and *I* values ranging from 30% to 40% were likely to occur. Despite being compared with the same historical mean *T*, we found that changes in the ranges of T_{diff} and *I* varied among regions.

Having identified the aforementioned relationships, we went on to assess changes in the *T* and *I* characteristics of each region. Table 2.1 shows the mean, 5th percentile,

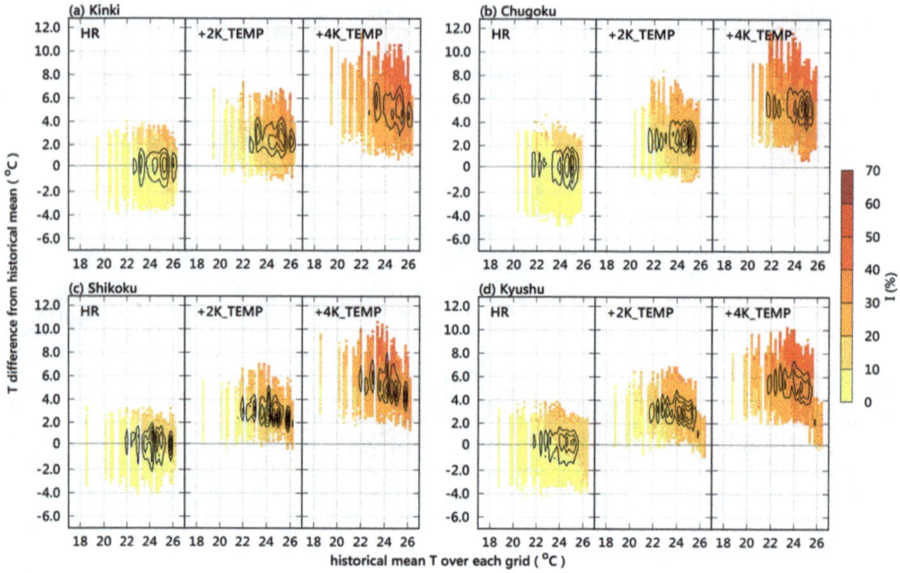

Fig. 2.3 The relationships between historical mean *T* over each grid, temperature difference from the historical mean *T* in each ensemble, and the percentage of chalky rice grains *I* in (**a**) Kinki, (**b**) Chugoku, (**c**) Shikoku, and (**d**) Kyushu. The contours indicate the relative counts of the data in 0.2% increments from 0.2%

and 95th percentile values for *T*, *S*, and *I* in each simulation. For Tohoku, the mean *T* in HR, +2K_TEMP, and +4K_TEMP simulations was estimated to be 21.7, 23.3, and 26.7 °C, respectively. Mean *I* was predicted to 6.7% and 21.6% for the +2K_TEMP and +4K_TEMP simulations, respectively. For Kanto, Hokuriku, and Tokai regions, the average and increasing trends of *T* and *I* were similar. The 90% probability interval range (from the 5th to 95th percentiles) for *T* and *I* was also approximately at the same level. Tosan is the coolest area in the study region, and for this region we determined that the mean *T* values for the HR, +2K_TEMP, and +4K_TEMP simulations were 20.4, 21.6, and 25.3 °C, respectively. The mean *I* for Tosan was also the smallest among those in the regions examined in this study, with values of 0.1%, 1.7%, and 9.3% for the HR, +2K_TEMP, and +4K_TEMP, simulations, respectively. For Kinki, Chugoku, Shikoku, and Kyushu regions, the mean *T* was larger than in other regions, resulting in correspondingly higher values of *I* than in the other regions. The average increase in *T* in these regions ranged from 2.6 to 2.9 °C between HR and +2K_TEMP, which was higher than the increases determined in other regions. This suggests that +2K_TEMP would have a large warming potential in regions that are currently warm. In response to a change from +2K_TEMP to +4K_TEMP, the Tohoku and Tosan regions are predicted to have the largest warming potential, with mean *T* increases of between 3.4 and 3.7 °C. This indicates that the temperature increase under +2K_TEMP to +4K_TEMP condition would be large in the currently cold regions. Although *I* was less than 10% in all

Table 2.1 Results of an impact assessment in each region. Mean values are shown, with 5th and 95th percentile values in parentheses

	T (°C)			*S* (MJ m^{-2} d^{-1})			*I* (%)		
	HR	+2K_TEMP	+4K_TEMP	HR	+2K_TEMP	+4K_TEMP	HR	+2K_TEMP	+4K_TEMP
Tohoku	21.7 (18.9, 24.4)	23.3 (19.4, 26.4)	26.7 (23.5, 29.5)	14.6 (11.5, 17.8)	14.2 (11.1, 17.6)	15.3 (12.1, 18.7)	1.0 (0.0, 7.3)	6.7 (0.0, 20.2)	21.6 (5.3, 36.2)
Kanto	23.1 (18.6, 26.3)	25.2 (19.4, 28.3)	27.8 (23.5, 30.9)	14.8 (11.1, 19.0)	15.1 (11.4, 19.4)	15.7 (12.0, 19.9)	6.7 (0.0, 21.0)	15.0 (0.0, 32.0)	27.2 (3.2, 46.5)
Hokuriku	23.7 (19.3, 26.5)	25.6 (19.9, 28.4)	28.3 (23.9, 30.8)	16.0 (12.7, 19.3)	15.9 (12.3, 19.4)	16.3 (12.9, 19.9)	5.2 (0.0, 17.6)	14.7 (0.0, 29.1)	28.4 (5.2, 42.8)
Tosan	20.4 (17.1, 23.6)	21.6 (16.8, 25.7)	25.3 (21.4, 28.8)	16.3 (12.7, 20.1)	15.7 (11.8, 20.0)	17.3 (13.0, 21.7)	0.1 (0.0, 0.0)	1.7 (0.0, 10.8)	9.3 (0.0, 26.4)
Tokai	23.3 (19.0, 26.7)	25.6 (19.9, 29.1)	28.1 (23.6, 31.4)	15.9 (12.2, 19.7)	16.6 (12.7, 20.9)	17.4 (13.3, 21.8)	6.3 (0.0, 21.2)	13.6 (0.0, 31.5)	24.7 (0.0, 44.2)
Kinki	24.2 (20.9, 26.7)	26.8 (23.7, 29.1)	29.4 (26.5, 31.8)	15.4 (12.0, 19.0)	16.2 (12.8, 19.7)	16.9 (13.3, 20.5)	8.7 (0.0, 21.6)	19.5 (4.2, 32.6)	33.4 (18.6, 46.9)
Chugoku	23.8 (20.7, 26.4)	26.6 (23.5, 28.9)	29.3 (26.6, 31.6)	15.6 (12.2, 19.1)	16.5 (13.0, 20.0)	17.0 (13.5, 20.6)	6.9 (0.0, 19.2)	17.9 (2.2, 30.7)	32.6 (18.6, 45.4)
Shikoku	23.8 (20.3, 26.4)	26.7 (23.5, 28.9)	29.2 (26.2, 31.6)	15.5 (11.9, 19.1)	17.0 (13.2, 20.8)	17.7 (13.8, 21.6)	7.4 (0.0, 19.7)	17.3 (1.8, 30.6)	30.0 (15.3, 44.7)
Kyushu	23.8 (20.4, 26.3)	26.7 (23.2, 28.7)	29.0 (25.9, 31.3)	15.3 (11.7, 18.9)	16.1 (12.5, 19.7)	16.4 (12.5, 20.2)	9.4 (0.0, 21.7)	19.2 (2.1, 32.2)	32.2 (16.3, 46.2)

regions based on the historical mean, in the +2K_TEMP simulation, the mean *I* exceeded 10% in seven regions, with the exceptions being Tohoku and Tosan. For the +4K_TEMP simulation, the mean *I* was predicted to exceed 20% in eight regions, with the only exception being Tosan. In these areas, the 95th percentile values may reach from 36.2% to 46.9%. Although T_{diff} and the change in *I* vary from region to region, there was a common tendency for CRGs to occur more frequently nationwide with increasing temperature.

Given that previous studies have not evaluated CRGs directly or used different GCM and greenhouse gas (GHG) emission scenarios, we are unable to make direct comparisons between the results of the present and previous studies. However, Okada et al. (2011) predicted that the percentage of first-grade rice would decrease in the Kyushu region with the continued progression of global warming. In addition, Ishigooka et al. (2017) indicated that the risk of rice quality deterioration would increase with the progression of global warming, and that the proportion of undamaged rice grains would decrease throughout Japan. The results of present study are consistent with these predictions.

However, it is conceivable that the results obtained in the present study are underestimates. According to recent experimental studies, increases in CO_2 concentration and soil temperature accelerate reductions in the percentage of undamaged rice in 'Koshihikari', with an estimated 13.7% decrease in the percentage of undamaged rice (Usui et al. 2016). These authors conducted experiments in an environment in which the CO_2 concentration was 200 μmol mol^{-1} higher than at present. In the +4K_TEMP meteorological simulation, the GHG concentration was set at 844.8 μmol mol^{-1} in 2090 according to representative concentration pathways 8.5 (Mizuta et al. 2017). Therefore, it is quite possible that a deterioration of rice grain quality will occur as CO_2 concentrations increase. This is because an increase in CO_2 concentration reduces leaf stomatal conductance, leading to a decrease in transpiration and a consequent increase in canopy temperature (Usui et al. 2014). To date, however, no models have been developed to estimate the occurrence of CRGs in relation to CO_2 concentrations. Accordingly, future studies should focus on developing such models.

2.3.2 Adaptation Simulation: Shifting the Transplanting Date

Figure 2.4 shows the changes in *T*, *S*, and *I* due to a shift in transplanting date under +2K_TEMP conditions. First, we show the changes in *T*, *S*, and *I* in each region when the transplanting date was shifted forward compared with the current transplanting date. For Tohoku and Tosan, both the 5th and 95th percentile values for *T* and *S* increased in all cases, although in each case the change in *I* tended to be small. For Kanto, Hokuriku, and Tokai, the 95th percentile value for *T* became lower and the 5th percentile value became higher, whereas the 95th percentile value for *S* decreased slightly as the transplanting date shifted forward. Consequently, although the change in the 95th percentile of *I* was small in these regions, it tended to decrease

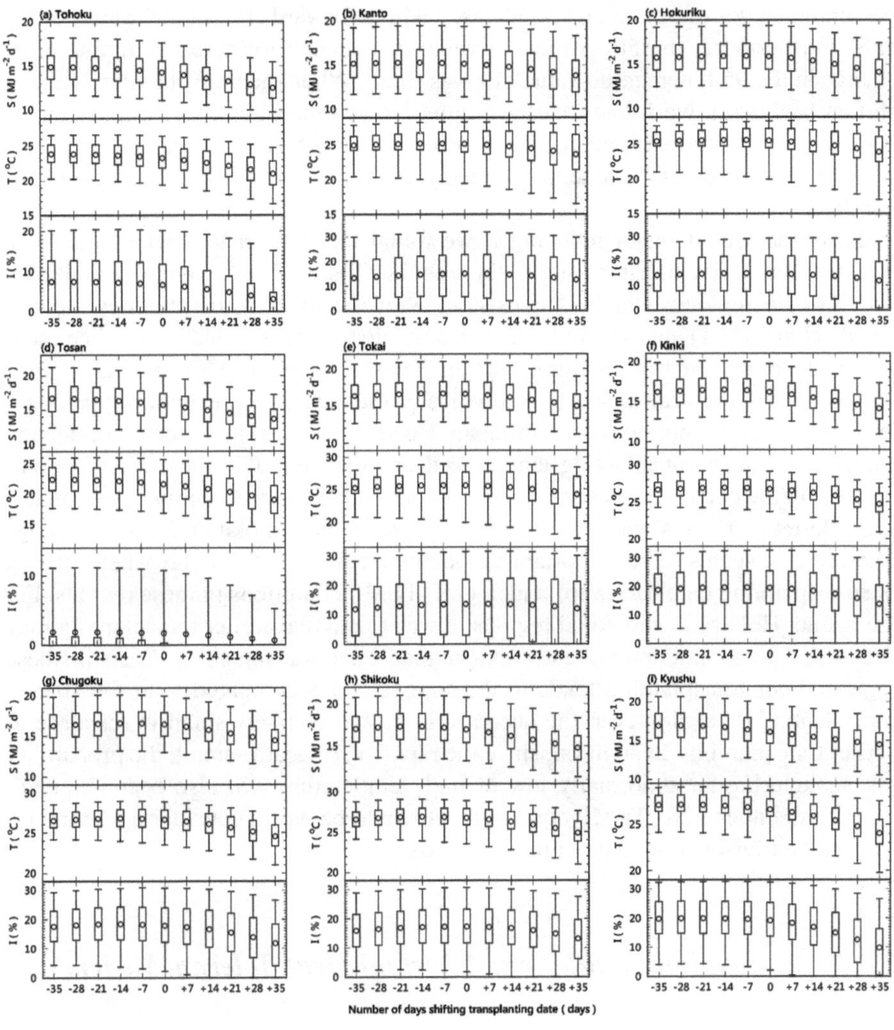

Fig. 2.4 The effect of shifting the transplanting date on *T*, *S*, and *I* under +2K_TEMP conditions in each region. The upper and lower whiskers indicate the 95th and 5th percentiles, respectively. The top and bottom boundaries of the boxes are the 75th and 25th percentiles, respectively. The circles in the boxes are the means

by shifting the transplanting date forward. For Kinki, Chugoku, and Shikoku, the 5th percentile value of *T* became higher as the transplanting date shifted forward. In contrast, the 95th percentile value of *T* showed no clear trend, although the value obtained by shifting the date 35 days forward was smaller than that for the current transplanting date. In case of *S*, the 5th and 95th percentiles showed no clear trend. The 5th and 95th percentile values of *I* tended to increase and decrease, respectively, with an increasingly earlier transplanting date. For Kyushu, the 5th and 95th percentile values of both T and S increased with a forward shift in the transplanting date. Furthermore, the 5th percentile value of I tended to increase, although the change in the 95th percentile value of I was small. When the transplanting date was shifted backward, we found that the results showed the same trend in all regions. Namely, the 5th and 95th percentile values of T and S tended to decrease and, as a consequence, those of I also tended to decrease with a backward shift in transplanting date.

From the viewpoint of reducing I, we found that late transplanting was more effective than early transplanting in the Tohoku, Tosan, Kinki, Chugoku, Shikoku, and Kyushu regions, which is consistent with the findings of previous studies. Okada et al. (2011) also showed that it is possible to suppress declines in rice quality by delaying the transplanting date in Kyushu, whereas Morita et al. (2016) remarked that later transplanting is a typical avoidance- and prevention-type countermeasure that is applied widely in Japan. However, the results of the present study showed that later transplanting tended to decrease S and T in these regions compared with the current transplanting date. In particular, the probability of encountering cold temperatures was expected to increase in the Tohoku and Tosan regions. Therefore, it seems to be important to maintain a sufficient T and S when shifting the transplanting date backward under +2K_TEMP conditions in these regions. For the Kanto, Hokuriku, and Tokai regions, it was found that earlier transplanting may decrease I to a similar or greater extent than later transplanting. Even in these regions, later transplanting tends to decrease T and S compared with the current transplanting date. However, the decrease in S was relatively small in response to earlier transplanting. In addition, since the range of T became small, the probability of encountering an abnormally low or high temperature will also become small. Therefore, under +2K_TEMP conditions, it would be worth considering a forward shift in the transplanting date in these regions.

2.3.3 Adaptation Simulation: A Virtual Heat-Tolerant Variety

The results of the adaptation simulation in which heat tolerance was virtually added to 'Koshihikari' under the +2K_TEMP conditions are shown in Fig. 2.5. In this section, we focus on the heat tolerance required to achieve the same level of I as in the impact assessment under the HR conditions, considering both mean and 95th percentile values (hereafter T_{req}). Although I was reduced in all regions by adding heat tolerance, T_{req} varied from region to region. For Tohoku, the difference in mean T between +2K_TEMP and HR was 1.6 °C (Table 2.1). However, it was found that a heat tolerance of 2.0 °C was necessary for T_{req}. For Hokuriku, the mean increase in T was 1.9 °C, but the required heat tolerance was 2.2 °C. In these regions, T_{req} was higher than the expected increase in T. One of the reasons for this discrepancy is considered to be solar radiation. The mean difference in S between +2K_TEMP and HR for Tohoku and Hokuriku was −0.4 and −0.1 MJ m^{-2} d^{-1}, respectively (Table 2.1). Because the model was developed by taking into account the

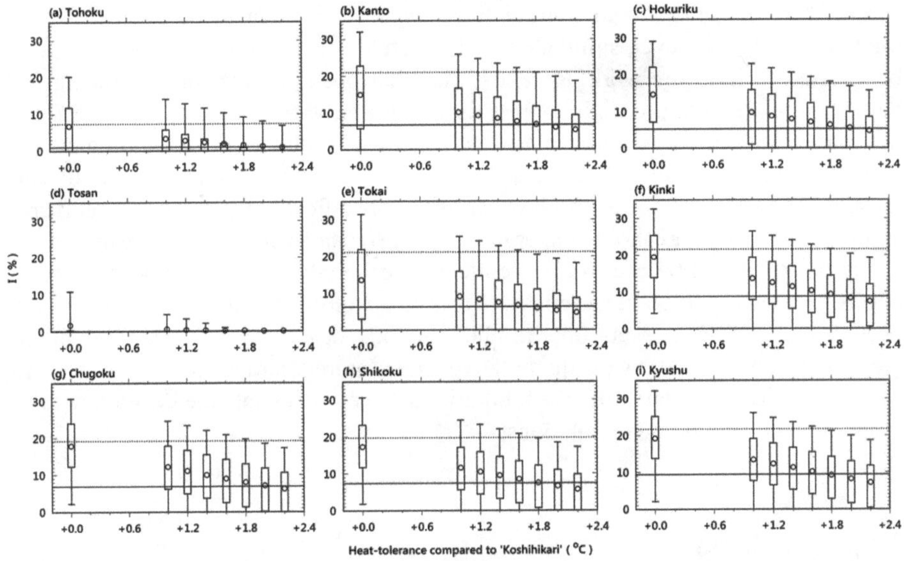

Fig. 2.5 Relationships between heat tolerance compared to 'Koshihikari' and *I* under +2K_TEMP conditions in each region. The upper and lower whiskers indicate the 95th and 5th percentiles, respectively. The top and bottom boundaries of the boxes are the 75th and 25th percentiles, respectively. The circles in the boxes are the means. Horizontal solid and dashed lines indicate mean and 95th percentile value of *I* in impact assessment under HR conditions, respectively. The boxplot at +0.0 is the same as in the impact assessment under the +2K_TEMP conditions

occurrence of milky-white grains, a type of CRG that changes depending on solar radiation, the temperature at which CRGs begin to occur becomes lower as solar radiation decreases in Eq. 2.1. Therefore, it is considered that enhanced heat tolerance greater than the actual temperature increase will be required in order to compensate for the decrease in solar radiation. For the Kanto, Tokai, Kinki, Chugoku, Shikoku, and Kyushu regions, the mean increases in T were 2.1, 2.3, 2.6, 2.8, 2.9, and 2.9 °C and the corresponding T_{req} values were 1.8, 2.0, 2.0, 2.0, 1.8, and 1.8 °C, respectively (Table 2.1). Thus, for these regions, it was found that T_{req} was lower than the expected increase in T. This is also considered to be attributable to solar radiation. In these regions, the subtraction of HR from +2K_TEMP in mean S was positive and ranged from +0.3 to +1.5 MJ m^{-2} d^{-1} (Table 2.1). On contrast to the aforementioned case, Eq. 2.1 is structured such that CRGs are less likely to occur due to an increase in solar radiation. In these regions, not only T but also S increased, which may have reduced T_{req}. For Tosan, we were unable to evaluate T_{req} because I was small even under the +2K_TEMP conditions. These results indicate that changes in solar radiation as well as temperature have a significant effect on T_{req}. The +2K_TEMP simulation projected that changes in temperature and solar radiation would vary according to region, and therefore when introducing heat-tolerant varieties, it is necessary to select varieties in accordance with the changes in temperature and solar radiation characterizing a particular region.

For this part of our study, simulations were carried out using a virtual heat-tolerant variety; however, simulations using actual heat-tolerant varieties must also be considered in future work. Recently, there has been progress in the breeding of new varieties resistant to high temperatures. Currently, the occurrence of CRGs in new varieties is often assessed relative to that in 'Koshihikari' (e.g. Kojima 2017); however, the heat tolerance characteristics of these new varieties compared with those of 'Koshihikari' are yet to be quantified. Specifically, the effects of temperature sensitivity and threshold temperatures on the occurrence of CRGs are unknown. In the future, it will be necessary to collect experimental data on the occurrence of CRGs in new heat-tolerant varieties and then to develop an estimation model for the occurrence of CRGs according to the methods described by Takimoto et al. (accepted). This will allow the quantitative temperature tolerance of new varieties to be assessed relative to that of 'Koshihikari', and thereby enable the development of more realistic adaptation simulations.

2.4 Conclusions

In this study, we evaluated the impact of meteorological conditions on the occurrence of CRGs in the rice cultivar 'Koshihikari' using a statistical model and a large number of ensemble climate projection data sets. As a consequence, we were able to predict that in response to global warming, the occurrence of CRGs would increase throughout Japan, although the response would vary from region to region. The most important outcome of our simulations is the provision of probability information on the occurrence of CRGs, based on year-to-year climate variability. We also considered and simulated two adaptations: shifting the transplanting date and using virtual heat-tolerant varieties. In both cases, we found that it would be important to select a suitable option that is responsive to future climate change in each region in order to reduce the occurrence of CRGs. This information will be useful to stakeholders such as farmers and policymakers when considering the impact of climate change on rice quality.

Acknowledgements We are grateful to Dr. Yasushi Ishigooka for providing the transplanting date data. Part of this work was supported by the Social Implementation Program on Climate Change Adaptation Technology (SI-CAT), MEXT, Japan. In this study, we utilized the database for Policy Decision-Making for Future Climate Change (d4PDF), which was produced using the Earth Simulator as a 'Strategic Project with Special Support' of JAMSTEC under corporation among the Program for Risk Information on Climate Change (SOUSEI), the SI-CAT, Integrated Research Program for Advancing Climate Models (TOUGOU), and the Data Integration and Analysis System (DIAS), which are all sponsored by the Ministry of Education, Culture, Sports, Science, and Technology (MEXT), Japan.

References

Cheng FM, Zhong LJ, Wang F et al (2005) Differences in cooking and eating properties between chalky and translucent parts in rice grains. Food Chem 90:39–46

Fukui S, Ishigooka Y, Kuwagata T et al (2015) A methodology for estimating phenological parameters of rice cultivars utilizing data from common variety trials. J Agric Meteorol 71:77–89

Geographical Survey Institute, Ministry of Land, Infrastructure, Transport, and Tourism of Japan (2006) Land use mesh 2006. http://nlftp.mlit.go.jp/ksj/gml/datalist/KsjTmplt-L03-a.html (in Japanese) Accessed 28 Nov 2018

Iizumi T, Yokozawa M, Nishimori M (2011) Probabilistic evaluation of climate change impacts on paddy rice productivity in Japan. Clim Chang 107:391–415

Imada Y, Maeda S, Watanabe M et al (2017) Recent enhanced seasonal temperature contrast in Japan from large ensemble high-resolution climate simulations. Atmosphere 8:57

Ishigooka Y, Kuwagata T, Nishimori M et al (2011) Spatial characterization of recent hot summers in Japan with agro-climatic indices related to rice production. J Agric Meteorol 67:209–224

Ishigooka Y, Fukui S, Hasegawa T et al (2017) Large-scale evaluation of the effects of adaptation to climate change by shifting transplanting date on rice production and quality in Japan. J Agric Meteorol 73:156–173

Kobayashi A, Hori K, Yamamoto T et al (2018) Koshihikari: a premium short-grain rice cultivar – its expansion and breeding in Japan. Rice. https://doi.org/10.1186/s12284-018-0207-4

Kojima Y (2017) Blanding of new rice cultivars to motivate farmers in Toyama. Hokuriku Crop Sci 55:65–66. (in Japanese)

Kondo M (2010) Effects of high-temperature stress on growth and grain filling in rice. Gamma F Symp 47:67–74

Kondo M, Morita S, Nagata K et al (2006) Effects of air temperature during ripening and grain protein contents on grain chalkiness in rice. Abstracts of Meeting of the CSSJ 222, 14 (in Japanese)

Masutomi Y, Takahashi K, Harasawa H et al (2009) Impact assessment of climate change on rice production in Asia in comprehensive consideration of process/parameter uncertainty in general circulation models. Agric Ecosyst Environ 131:281–291

Masutomi Y, Arakawa M, Minoda T et al (2015) Critical air temperature and sensitivity of the incidence of chalky rice kernels for the rice cultivar “Sai-no-kagayaki”. Agric For Meteorol 203:11–16

Ministry of Agriculture, Forestry and Fisheries (MAFF) (2018) Rice. http://www.hinshu2.maff.go.jp/info/sinsakijun/kijun/1440.pdf (in Japanese). Accessed 11 Mar 2019

Ministry of Agriculture, Forestry, and Fisheries of Japan (MAFF) (2009). http://www.maff.go.jp/j/tokei/kouhyou/beikoku_ryutu/pdf/suito_hinsyu_09.pdf (in Japanese). Accessed 28 Nov 2018

Ministry of Agriculture, Forestry, and Fisheries of Japan (MAFF) (2018) Statistics of agriculture, forestry, and fisheries. http://www.maff.go.jp/j/tokei/kouhyou/sakumotu/menseki/attach/pdf/index-22.pdf (in Japanese). Accessed 28 Nov 2018

Mizuta R, Murata A, Ishii M et al (2017) Over 5,000 years of ensemble future climate simulations by 60-km global and 20-km regional atmospheric models. Bull Am Meteorol Soc 98:1383–1398

Morita S (2005) The occurrences of immature grain with white portions and deep ditch, and grain weight decrease in rice under high temperature during ripening. J Agric Sci 60:442–446. (in Japanese)

Morita S (2008) Prospect for developing measures to prevent high-temperature damage to rice grain ripening. Japanese J Crop Sci 77:1–12. (in Japanese)

Morita S, Wada H, Matue Y (2016) Countermeasures for heat damage in rice grain quality under climate change. Plant Prod Sci 19:1–11

Ohno H, Sasaki K, Ohara G et al (2016) Development of grid square air temperature and precipitation data compiled from observed, forecasted, and climatic normal data. Clim Biosph 16:71–79. in Japanese

Okada M, Iizumi T, Hayashi Y et al (2011) Projecting climate change impacts both on rice quality and yield in Japan. J Agric Meteorol 67:285–295
Sasahara H, Goto A, Shigemune A et al (2018) A new variety for sushi "Eminokizuna". Bull NARO Agric Res Cent Reg 5:1–18. (in Japanese)
Seino H (1997) Global climate change and food production in Japan. J Agric Meteorol 52:367–376
Takimoto T, Masutomi Y, Tamura M et al (accepted) The effect of air temperature and solar radiation on the occurrence of chalky rice grains in rice cultivars "Koshihikari" and "Akitakomachi". J Agric Meteorol
Terashima K, Saito Y, Sakai N et al (2001) Effects of high air temperature in summer of 1999 on ripening and grain quality of rice. Japanese J Crop Sci 70:449–458. in Japanese
Usui Y, Sakai H, Tokida T et al (2014) Heat-tolerant rice cultivars retain grain appearance quality under free-air CO_2 enrichment. Rice 7:6
Usui Y, Sakai H, Tokida T et al (2016) Rice grain yield and quality responses to free-air CO_2 enrichment combined with soil and water warming. Glob Chang Biol 22:1256–1270
Wakamatsu K, Sasaki O, Uezono I et al (2007) Effects of high air temperature during the ripening period on the grain quality of rice in warm regions of Japan. Japanese J Crop Sci 76:71–78. (in Japanese)
Wakamatsu K, Sasaki O, Uezono I et al (2008) Effect of the amount of nitrogen application on occurrence of white-back kernels during ripening of rice under high- temperature conditions. Japanese J Crop Sci 70:424–433. (in Japanese)
Wang J, Wan X, Li H et al (2007) Application of identified QTL-marker associations in rice quality improvement through a design-breeding approach. Theor Appl Genet 115:87–100
Yamakawa H, Hirose T, Kuroda M et al (2007) Comprehensive expression profiling of rice grain filling-related genes under high temperature using DNA microarray. Plant Physiol 144:258–277
Yoshida K, Sugi M, Mizuta R et al (2017) Future changes in tropical cyclone activity in high-resolution large-ensemble simulations. Geophys Res Lett 44:9910–9917
Zhang S, Tao F (2013) Modeling the response of rice phenology to climate change and variability in different climatic zones: comparisons of five models. Eur J Agron 45:165–176

Chapter 3
New Approaches Combined with Environmental Control for Enhancing Heat-Tolerant Rice Breeding in Japan

Hiroshi Wada

Abstract Reducing the occurrence of heat-related damages, such as rice chalkiness and spikelet sterility, is the major goal in global rice production under climate change. In Japan, multiple heat-tolerant cultivars have been developed to reduce the occurrence of chalky rice using conventional breeding, whereas at present no cultivars have been developed as heat-tolerant cultivars for spikelet sterility. Numerous studies have investigated the cause(s) of each phenomenon mostly at tissue level, while the exact mechanism(s) behind each phenomenon at cell level is still not well understood. Considering the predicted increase in heat risk, development of superior heat-tolerant cultivars is strongly desired as a countermeasure. Heat tolerance of multiple lines/cultivars has been evaluated under fluctuating environmental conditions; however, field evaluation is demanding and complicated because of the low reproducibility of heat conditions in the field. To address these issues, robust high-throughput screening methods are required. Recently, several attempts to overcome the heat-related damages have been made by using environmental control. In this chapter, recent progress focused on these attempts and future prospects have been described.

3.1 Background and Status of Rice Breeding on Heat Tolerance in Japan

The frequency and intensity of elevated global temperature and dryness are likely to increase in eastern Asia (IPCC 2007). In rice production, chalkiness and spikelet sterility are two major heat-related damages that prevent the establishment of steady rice production in the area, including Japan (see the general reviews, Jagadish et al. 2015; Morita et al. 2016; Ishimaru et al. 2016). Rice chalkiness often degrades the appearance and milling quality of rice, whereas spikelet sterility reduces rice yield.

H. Wada (✉)
Kyushu Okinawa Agricultural Research Center, National Agriculture and Food Research Organization, Chikugo, Fukuoka, Japan
e-mail: hwada@affrc.go.jp

T. Iizumi et al. (eds.), *Adaptation to Climate Change in Agriculture*,
https://doi.org/10.1007/978-981-13-9235-1_3

Fig. 3.1 Images of perfect rice and five types of chalky rice induced by heat-related damages. (Photographs courtesy of Dr. Hiroshi Nakano)

Chalky rice is often caused by heat (Tashiro and Wardlaw 1991), low light (Nagato and Chaudhry 1970), and foehn-like dry wind conditions (Ishihara et al. 2005; Wada et al. 2011, 2014) during the ripening stage. There are multiple types of chalky rice, such as white-belly rice, white-back rice, basal-white rice, milky-white rice, and white-core rice (Hoshikawa 1989; see Fig. 3.1), categorized by the location of chalky region formed in a part of endosperm. A statistical method that links phenotypic data and genotypic data established in the 1980s, known as quantitative trait loci (QTL) analysis, has been frequently used in rice breeding to understand the genetic basis for most agronomically important traits including rice chalkiness. Several effective QTLs associated with rice chalkiness have been identified during the last decade (e.g., Kobayashi et al. 2013 for white-back rice). Marker-assisted breeding for quality improvement is currently the main selection approach for developing heat-tolerant cultivars in rice breeding.

A leading cultivar, 'Hinohikari', widely cultivated for approximately the last 30 years in the flat areas of the western Japan, is susceptible to high temperature during ripening. The occurrences of chalky grains in 'Hinohikari' have been observed more frequently for the past 15 years. Figure 3.2 shows the percentage of the first-grade rice by grain inspection in the extremely hot year, 2010, and the 10-year mean between 2007 and 2016 of each prefecture as well as the national average in Japan. The 10-year mean showed severe rice quality deterioration

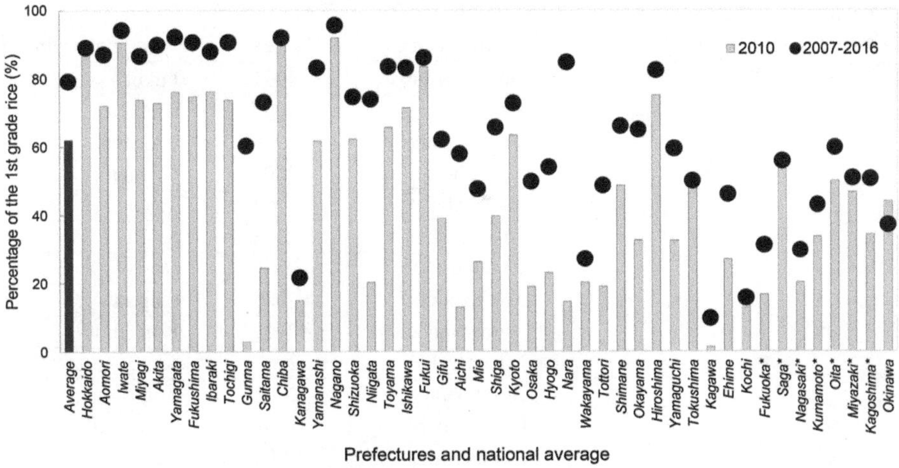

Fig. 3.2 Percentage of the first-grade rice by inspection in 2010 and the 10-year mean between 2007 and 2016 of prefectures and the national average (leftmost shown in black) in Japan. Data were obtained from MAFF. Asterisk indicates the prefectures in Kyushu region

observed in most prefectures of south-western Japan, including Kyushu region (see seven prefectures from Fukuoka to Kagoshima in Fig. 3.2). Except for Hokkaido and Okinawa prefecture, there was a remarkable quality deterioration essentially in all prefectures in 2010, which was due to the extremely high temperatures recorded for the long duration during the ripening. To reduce heat risk during the ripening stage in the warmer climate predicted for the future, breeding strategies in south-western Japan have mainly focused on the development of heat-tolerant cultivars, as well as shade tolerance from typhoons that often hit the region during the summer.

A heat-tolerant cultivar with a superior palatability, 'Nikomaru,' was developed in 2005 at NARO Kyushu Okinawa Agricultural Research Center. 'Nikomaru' exhibited much less chalky rice than 'Hinohikari' when exposed to high temperature plus low light intensity during ripening (Sakai et al. 2007). Since then, 'Nikomaru' has been gradually spreading in several regions including Kyushu area. It has been shown that 'Nikomaru' accumulated greater amount of stem nonstructural carbohydrates at heading than 'Hinohikari' (Morita and Nakano 2011), suggesting that assimilates would be supplied from the temporarily stored stems to the developing grains, even when assimilates supply from leaf photosynthesis declined under heat and/or low light conditions. Two brand new heat-tolerant cultivars, 'Minoritsukushi' in Fukuoka prefecture (Wada et al. 2016) and 'Natsuhonoka' in Kagoshima prefecture (Wakamatsu et al. 2016), developed from 'Nikomaru' as a parent, are currently spreading in each prefecture. Besides Kyushu region, many heat-tolerant cultivars have been developed in each region in Japan (Table 3.1).

Based on growth chamber experiments, it is generally accepted that exposure to temperatures higher than 35 °C at flowering induces spikelet sterility, leading to yield loss (Satake and Yoshida 1978). At present, deterioration of rice quality caused

Table 3.1 List of heat-tolerant rice cultivars during ripening stage developed in Japan

Cultivar name	Year of release	Breeding origin
Hanaechizen[a]	1991	Fukui
Fusaotome[a]	1997	Chiba
Yumemizuho	1999	Ishikawa
Koshiibuki	2000	Niigata
Tentakaku	2003	Toyama
Nikomaru	2005	NARO
Kinumusume	2005	NARO
Yukinkomai	2005	Niigata
Kumasannochikara	2007	Kumamoto
Mineharuka[a]	2007	Aichi
Akihonami	2008	Kagoshima
Genkitsukushi	2009	Fukuoka
Sagabiyori	2009	Saga
Tsuyahime[a]	2009	Yamagata
Oidemai	2011	Kagawa
Eminokizuna[a]	2011	NARO
Harumoni	2011	NARO
Otentosodachi[a]	2011	Miyazaki
Tochiginohoshi[a]	2011	Tochigi
Mie23	2011	Mie
Sainokizuna	2012	Saitama
Koinoyokan	2014	NARO
Minoritsukushi	2015	Fukuoka
Natsuhonoka[a]	2015	Kagoshima
Shinnosuke	2015	Niigata
Kumasannokagayaki	2016	Kumamoto
Ichihomare	2017	Fukui

NARO: National Agriculture and Food Research Organization, Japan

[a]Cultivars chosen standard cultivars for heat tolerance during ripening stage (MAFF 2018)

by heat stress during ripening is a key issue in Japan (Morita et al. 2016; Ishimaru et al. 2016). However, spikelet sterility is expected to become more serious due to increased carbon dioxide in the atmosphere under high temperature conditions at flowering (Horie et al. 1996). In effect, 20% of severe heat-induced rice spikelet sterility was observed in the summer of 2007 in the central region of Japan (Hasegawa et al. 2011). To mitigate heat-induced spikelet sterility, robust breeding strategies are required. Some investigators have been working on incorporating early flowering traits into certain cultivars to avoid heat risk at flowering (see recent reviews by Jagadish et al. (2015) and Ishimaru et al. (2016)). The shift in flower opening time to a period with cooler temperature has been demonstrated to be an effective way of avoiding heat risk at flowering.

Satake and Yoshida (1978) showed that heat-induced spikelet sterility is caused by heat exposure of panicles during booting and flowering. The effects of heat stress at flowering stage seem to have received the most attention, while little is known about the potential influence of heat history on spikelet sterility. However, heat duration often lasts for more than a few days under field conditions. Therefore, the potential effects of the history of heat conditions prior to heading should be measured. In addition to the introduction of early-morning flowering phenotypes, development of superior heat-tolerant cultivars is urgently desired. To date, it seems likely that no reliable heat-tolerant screening method has been used; however, such a screening method has been available for use in combination with the environmental control, as described below (see Sect. 3.2.1).

Rice chalkiness is an endosperm phenotype with cell heterogeneity that was formed under adverse conditions as described above. It is well known that white-back rice and basal-white rice (see Fig. 3.1) are often formed when exposed to heat at low plant nitrogen level. And, the occurrence of this chalky rice is suppressed by increasing level of the nitrogen using the nitrogen fertilizer for top dressing, exhibiting substantial varietal difference (e.g., Wakamatsu et al. 2008 for white-back rice). Based on the finding, there may be some relationships between cultivar and cell-specific heat responses associated with internal nitrogen level. Numerous comprehensive studies using molecular approaches have been conducted during the last two decades; however, our knowledge on what exactly occurs in the endosperm cells to reduce chalkiness when applying nitrogen prior to heat conditions has not improved. More recently, the direct assay of the endosperm cellular metabolisms, called '*on-site cell-specific analysis*,' has become possible under heat conditions (see Sect. 3.2.2). The underlying mechanism(s) and varietal differences for nitrogen-induced suppression of white-back rice under heat conditions are currently being investigated, using this system. Several new attempts using this direct assay in combination with environmental control may improve rice-breeding programs. The background and recent progress focused on these attempts and future prospects are discussed.

3.2 Use of Control Environments to Accelerate Crop Breeding

3.2.1 Heat Tolerance Assay System for Rice Spikelet Sterility

High temperatures, above about 35 °C, markedly reduce rice yield, due to spikelet sterility (Matsui et al. 2001). By using the weather data for the 30 years between 1971 and 2000 and four climate scenarios with doubled CO_2 concentration until 2090s in the rice growth simulation model SIMRIW (Horie 1995), Nakagawa et al. (2003) predicted that an increase in heat tolerance of approximately 1.5 °C at flowering would greatly reduce the incidence of spikelet sterility in central and southern

Japan. In addition to heat itself, increased humidity under heat conditions raises the risk of spikelet sterility (Matsui et al. 1997a; Weerakoon et al. 2008). Wind speed above 0.85 m s^{-1} decreases spikelet fertility at 37.5 °C due to the reduction in the pollen grain number shed on stigma (Matsui et al. 1997b). High carbon dioxide concentration also affects sterility, presumably due to the increased panicle temperature through the stomatal closure and reduced transpiration (Matsui et al. 1997a). Although the promotion of breeding heat-tolerant rice cultivars is an urgent task, evaluating the heat tolerance among numerous lines/cultivars has been challenging in breeding programs.

Varietal differences in heat-induced spikelet sterility have been reported in previous studies (Matsui et al. 1997a, 2001; Matsui and Omasa 2002; Prasad et al. 2006; Jagadish et al. 2007, 2008; Jagadish et al. 2010; Maruyama et al. 2013; Hakata et al. 2017). Due to these studies being conducted under various conditions in different compartments, such as temperature-gradient chambers, growth chambers, and greenhouses, there was some inconsistency in the data, as described by Hakata et al. (2017). It has been known that rice panicle is a heat-sensitive organ (Nishiyama and Satake 1981), particularly at nighttime (Morita et al. 2004). When conducting the growth chamber analysis, it is anticipated that the distances between panicles and lamps as a 'heat source' might be inconsistent due to the varietal difference in the plant height, leading to the various thermal environments at the top panicle position.

Hakata et al. (2017) have recently developed a new custom-built assay system that allows researchers to examine the varietal differences in heat-induced spikelet sterility with high reproducibility. In this assay system, an artificial paddy that can contain a maximum of 216 pots was placed in each of the two environmentally controlled growth chambers (Fig. 3.3). In this system, the panicle-top position of rice plants grown in the artificial paddy can be adjusted, so that height uniformity is ensured in all cultivars/lines during heat conditions. The air temperature and relative humidity in the growth chambers were controlled by using the sensors placed at the same panicle height of the plants in the chambers. Using this system, the authors revisited the question for the varietal differences of heat tolerance with the heat-tolerant rice cultivars previously reported. Furthermore, Hakata et al. conducted a complete set of screenings with a total of 118 genotypes from the Rice Core Collection available at the NARO Genebank, in addition to two cultivars, 'N22' already recognized as a heat-tolerant cultivar (e.g., Li et al. 2015) and 'Koshihikari' as a reference genotype. From these results, several genotypes that exhibit much greater heat tolerance than 'N22' have been successfully identified (Hakata et al. unpublished data). This system allows researchers to examine heat tolerance of multiple cultivars at a set temperature throughout the year in the isolated chambers. Although confirmatory analysis will be required under field conditions, this system has shown promise. The system could be used for the evaluation of not only spikelet sterility, but also rice chalkiness with multiple cultivars. It is also anticipated that using this approach would be practically applicable to most crop breeding.

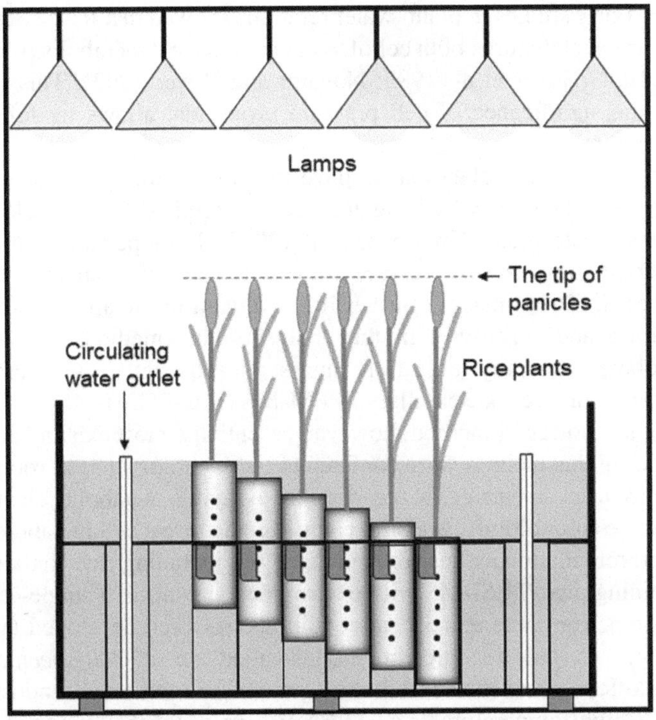

Fig. 3.3 Schematic diagram of the newly developed heat-tolerant assay system with an artificial rice paddy in the growth chambers. When exposed to heat, the position of height-adjustable pots was individually adjusted, so that the distance between the metal halide lamps as a heat source and the tip of panicles could be uniform for all cultivars/lines. The system could be useful for selecting heat-tolerant varieties, eliminating the potential artifacts. After Hakata et al. (2017)

3.2.2 Analysis of Rice Chalkiness Using On-Site Cell-Specific Analysis

Cell heterogeneity has been increasingly recognized as a biologically important phenomenon observable in most living organisms. Chalk-like appearance formed in the rice endosperm is a typical phenotype of cell heterogeneity. Currently, multi-omics analyses on heat-related damages, such as chalkiness and spikelet sterility, have been actively made at kernel and tissue level. Even with tremendous efforts, the exact cellular mechanism(s) behind each phenomenon and their varietal differences have not been fully understood due to the limitation of resolution. To assay heat-related metabolic changes at cell level, it is necessary to develop some robust analytical method performable in the target cells of the kernels growing under heat conditions.

A cell pressure probe technique (Hüsken et al. 1978), known to be the only device available for directly assaying turgor pressure in plant cells, has been long used for a number of studies in plant–water relations (Steudle 1993; Tomos and

Leigh 1999). Early studies in plant–water relations showed that tissue excision prior to any measurement disturbs both cellular water status and metabolism, particularly in growing cells (Boyer et al. 1985; Nonami and Boyer 1993). This observation highlighted the significance of cell pressure probe that allows to directly use in intact plant organs.

Recently, single-cell metabolomics have made revolutionary progress in plant sciences (e.g., see the review by Yang et al. 2017) mainly due to the addition of the Orbitrap mass spectrometer. Gholipour et al. (2013) developed an in situ analytical method by combining the cell pressure probe with an Orbitrap mass spectrometer. Nakashima et al. (2016) made a further improvement in the apparatus to increase both resolution and sensitivity in the analysis. This method, called 'picolitre pressure-probe-electrospray-ionization mass spectrometry (picoPPESI-MS)', referred as 'internal electrode capillary PPESI-MS (IEC-PPESI-MS)' in Nakashima et al. (2016), is a powerful method; however, essentially, most existing cell metabolomics including this method were confined to the laboratory use at room temperature. This has been a barrier in the case of assaying metabolic changes under temperature stress conditions. And, there has been no direct cell metabolome analysis conducted relating to any heat-related damages, including rice chalkiness.

By combining picoPPESI-MS with environmental control, a single-cell metabolome analysis performable under heat conditions has been developed (Wada et al. 2019; see Fig. 3.4). This analytical method is called '*on-site cell-specific analysis*.' This system offers a real-time metabolites profiling in the target endosperm cells growing at a high temperature (e.g., at 34 °C) prior to chalky formation. More recently, the cellular approach using this '*on-site cell-specific analysis*' has been applied for directly assaying the metabolic changes in developing rice endosperms prior to chalky formation to further extend our knowledge of the cellular mechanisms on chalky (white-back rice) formation and nitrogen-associated suppression effect.

Regarding the rice chalky formation, it has been known that the air spaces formed between loosely packed starch granules cause random light reflection that turns into chalkiness (Tashiro and Wardlaw 1991). Because starch is the most abundant component in rice endosperms, high temperature-related studies on grain chalkiness have been mostly focused on starch biosynthesis at the molecular level (see the review by Sreenivasulu et al. 2015). It is true that partial arrest of amyloplast development leads to chalky formation, although the source organelle(s) of the air spaces remained in a part of the endosperm cells have not been clearly identified presumably because of the technical difficulty when conducting a transmission electron microscopy (TEM) in mature kernels (see Discussion in Hatakeyama et al. 2018).

In recent years, there is accumulating evidence that vacuole-like structures, such as vesicles and vacuoles, remaining in cytosol may also participate in the air space formation in chalky cells (see Li et al. 2014 for white belly and white core rice; Wada et al. 2019 for white-back rice; Hatakeyama et al. 2018 for milky-white rice), besides the declined amyloplast development at high temperature. Under normal growth conditions, the packing density of amyloplasts in endosperm cells increased as the volume occupied with other organelles including vacuole-like structures decreased to fill up the endosperm cells at maturation (Hoshikawa 1989;

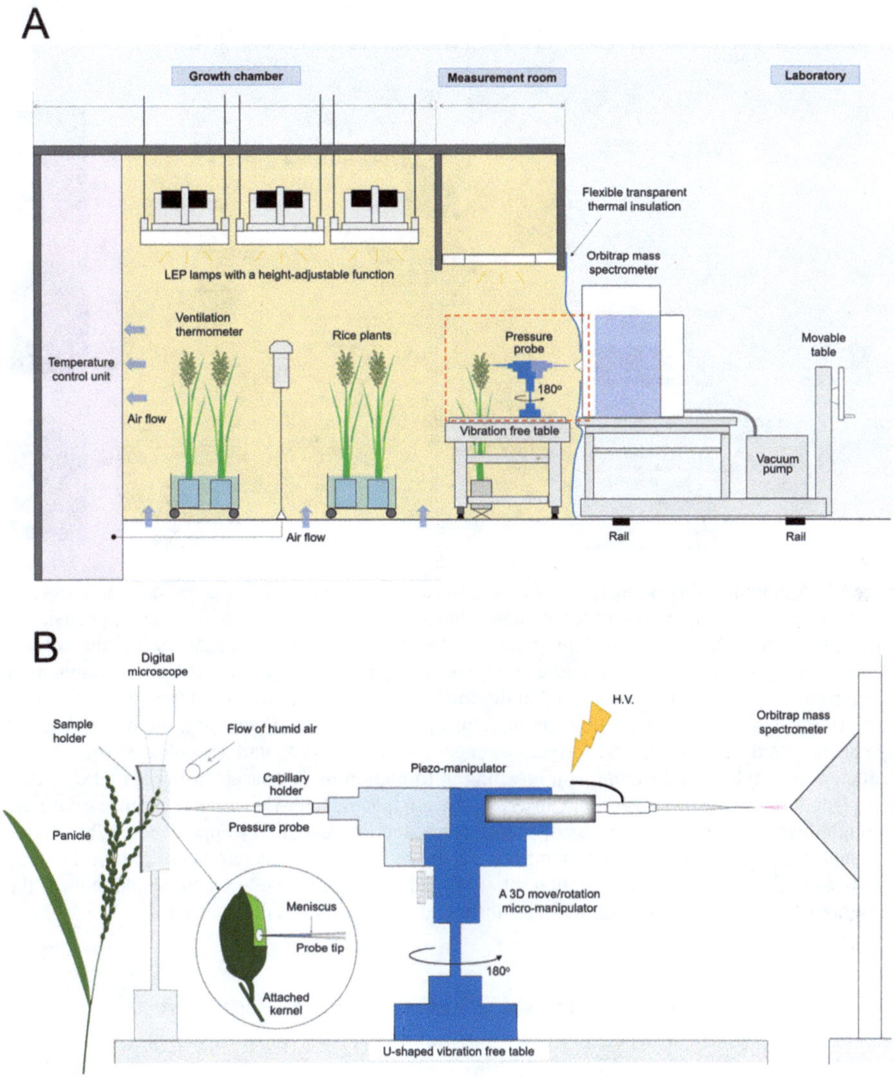

Fig. 3.4 Diagram of the side view of the system of '*on-site cell-specific analysis*'. (**a**) The entire diagram of the system placed in the laboratory and (**b**) expanded figure of picolitre pressure-probe-electrospray-ionization mass spectrometry (picoPPESI-MS) shown in red rectangle in **a**. The system is composed of picoPPESI-MS and environmental control. Using this system, metabolome analysis was conducted in single endosperm cells of intact seeds produced under controlled environment conditions. In the developing rice endosperms under humid conditions, a part of hull and pericarp tissue was removed. By utilizing a cell pressure probe, the probe tip was then introduced into the target endosperm cells, where the high frequency of chalkiness is expected to form at maturation, and the cellular fluid discharged in the quartz capillary was collected. The Piezo-manipulator was rotated 180° and the metabolites in the fluid were instantly analyzed in Orbitrap mass spectrometer by applying a high voltage (**b**). (Adopted from Wada et al. 2019)

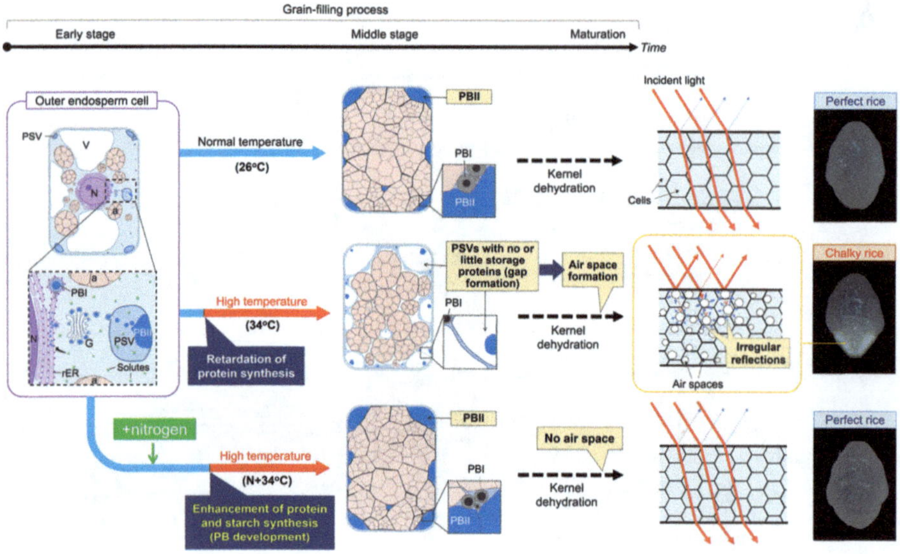

Fig. 3.5 A diagram illustrating the proposed process of heat-induced formation of chalkiness and nitrogen-enhanced adaptation in rice kernels. Under normal conditions, as the packing density of amyloplasts increases in the developing outer endosperm cells, there is a reduction in the volume of the cytosol, which is mainly occupied with vacuoles (V), resulting in adequate development of amyloplasts and protein bodies (PBs) that diminish the cytosolic space. When the nitrogen level is low prior to high temperature, the osmotic adjustment in the cells is less and the rate of protein synthesis is reduced to preserve protein storage vacuoles (PSVs) and vacuoles, with a partial reduction in amyloplast development, resulting in formation of air spaces during kernel dehydration. Due to the irregular reflection caused by the air spaces, light is scattered and does not pass through the kernel, leading to a loss of transparency and hence a chalky appearance. When nitrogen supply is adequate prior to high temperature, the cells maintain the rate of protein synthesis by strong osmotic adjustment and sustain normal PB and amyloplast development, which reduces the presence of air spaces, thus resulting in a substantial reduction in chalky kernels

Del Rosario et al. 1968). In contrast to the endosperm under no-stress conditions, the behavior of these vacuolar compartmentations in the cytosol is likely to be altered in the cells when exposed to stress conditions including heat, together with the partial decline in starch biosynthesis.

During foehn-induced chalky ring formation, known as milky-white rice, it has been shown that the preservation of numerous vacuole-like compartments including protein storage vacuoles in cytosol could be the main cause of air space formation in the inner endosperms (Hatakeyama et al. 2018). For heat-induced white-back rice formation (see Fig. 3.5), it has been reported that heat disrupted amyloplast and protein body development, and vacuolar compartments, mainly protein storage vacuoles, were expanding over time and preserved in the cytosol of outer endosperm cells of dorsal side of kernels, turning into air spaces that lead to white-back rice (Wada et al. 2019). In contrast, endosperm cells in plants supplied with nitrogen sustained both protein body formation and amyloplast development, which allowed them to suppress the formation of chalkiness even under high-temperature conditions (Fig. 3.5). In addition to the reduction in cell turgor under heat conditions, it should be emphasized that treatment differences in cell metabolites, particularly redox metabolites including cysteine, ascorbic acid, and glutathione, were cell-specifically identified prior to the changes in protein body morphology by combining '*on-site cell-specific analysis*' with time-course TEM analysis (Wada et al. 2019).

Different from the tissue-level analysis, the use of pinpoint cell-omics analysis may help to identify the control gene(s) in the target cells through better understanding of the cellular mechanisms. Keeping this in mind, further analysis of heat-related damages, such as varietal differences in the occurrence of white-back rice associated with nitrogen application, kernel growth inhibition at high night temperature (Morita et al. 2005), and varietal differences in heat-induced spikelet sterility, may provide useful information in rice breeding.

3.3 Conclusions and Perspectives

As described previously, heat-tolerant cultivars have been developed to reduce the number of chalky rice under extreme heat conditions. According to the latest standard evaluation for heat-tolerant cultivars during ripening stage, recently reported in the Ministry of Agriculture, Forestry, and Fisheries (MAFF), Japan (Sato et al. 2018; MAFF 2018), 'Nikomaru' has been rated as 'moderately heat tolerant' among tolerant cultivars in the south-west region in Japan. In Japan, cultivars such as 'Fusaotome', 'Tsuyahime', 'Eminokizuna', 'Otentosodachi', 'Tochiginohoshi', and 'Natsuhonoka' have been classified as highly tolerant to heat during the ripening stage (see asterisks in Table 3.1). The data implies that 'Natsuhonoka' might have enhanced heat tolerance inherited from a parent cultivar, 'Nikomaru' (Sato et al. 2018; MAFF 2018). Given the finding that elevated CO_2 would reduce the threshold air temperature for heat stress (Usui et al. 2014), development of superior heat tolerance should be promoted to overcome more severe heat risk during the ripening stage.

Crop breeding remains a time-consuming and complicated process. However, it is ideal to shorten the generation cycle to accelerate breeding for developing new cultivars. More recently, Watson et al. (2018) reported a powerful method that shortens the generation cycle by extending the photoperiod using supplemental LED lightning in controlled environments. Although this method worked well for the long-day crops reported, the direct application of the speed breeding protocols to short-day plants, such as rice or maize, may not be successful, as Watson et al. (2018) pointed out. Considering the various temperature/photoperiod sensitivities among lines/cultivars, some optimization will be required in rice to accelerate generation cycle. A similar use of environmental control may be applicable to other crops.

Development of some robust high-throughput screening methods, particularly in heat-induced spikelet sterility, appears to be an important step for the selection process. Heat tolerance assay system combined with the environmental control (Hakata et al. 2017, see Sect. 3.2.1) introduced here would be a powerful approach for screening rice spikelet sterility. Understanding the exact mechanism(s) and identifying varietal differences have become more important for the improvement of cultivation methods as well as breeding. Identifying the varietal differences in cellular heat responses associated with nitrogen level is currently under investigation. Further analysis may shed additional light on enhancing heat tolerance through the identification of the regulatory gene(s) to high-temperature conditions.

Finally, targeted mutagenesis technologies, such as genome editing, are promising approaches to accelerate breeding programs in rice. The main target with this technique in Japan is directed mostly toward generating desirable mutations for high-yielding rice, not for heat tolerance. At present, the combination of conventional breeding and marker-assisted breeding focusing on several QTLs is the main approach for the development of heat-tolerant cultivars for rice quality improvement. Even with these tremendous efforts, available QTLs/genes are still limited for rice breeding. It should not be ignored that the predicted rise in temperature can potentially lead to other risks, including pests and diseases in the field, depending on the regions and climate conditions. Some of the released heat-tolerant cultivars are known to lack resistance to some pests or diseases. From a practical viewpoint, further improvement of rice cultivars under heat conditions will need to be achieved by pyramiding with other important traits, such as biotic tolerances and good palatability, in addition to heat tolerance.

Acknowledgments The authors gratefully acknowledge Drs. Hiroshi Nakano, Makoto Hakata, and Yuto Hatakeyama for helpful comments on the manuscript. This work was supported by JSPS KAKENHI Grant Number 16H02533 and 16H04870.

References

Boyer JS, Cavalieri AJ, Schulze ED (1985) Control of the rate of cell enlargement – excision, wall relaxation, and growth-induced water potentials. Planta 163(4):527–543

Del Rosario AR, Briones VP, Vidal AJ, Juliano BO (1968) Composition and endosperm structure of developing and mature rice kernel. Cereal Chem 45(3):225–235

Gholipour Y, Erra-Balsells R, Hiraoka K, Nonami H (2013) Living cell manipulation, manageable sampling, and shotgun picoliter electrospray mass spectrometry for profiling metabolites. Anal Biochem 433(1):70–78

Hakata M, Wada H, Masumoto-Kubo C, Tanaka R, Sato H, Morita S (2017) Development of a new heat tolerance assay system for rice spikelet sterility. Plant Methods 13:1–8

Hasegawa T, Ishimaru T, Kondo M, Kuwagata T, Yoshimoto M, Fukuoka M (2011) Spikelet sterility of rice observed in the record hot summer of 2007 and the factors associated with its variation. J Agric Meteorol 67(4):225–232. https://doi.org/10.2480/agrmet.67.4.3

Hatakeyama Y, Masumoto-Kubo C, Nonami H, Morita S, Hiraoka K, Onda Y, Nakashima T, Nakano H, Wada H (2018) Evidence for preservation of vacuolar compartments during foehn-induced chalky ring formation of *Oryza sativa* L. Planta 248:1263–1275. https://doi.org/10.1007/s00425-018-2975-x

Horie T (1995) Rice production in Japan under current and future climates. In: Modeling the impact of climate change on rice production in Asia. CAB International, Wallingford, pp 143–164

Horie T, Matsui T, Nakagawa H, Omasa K (1996) Effects of elevated CO_2 and global climate change on rice yield in Japan. In: Climate change and plants in East Asia. Springer-verlag, Tokyo

Hoshikawa K (1989) The growing rice plant: an anatomical monograph. Nobunkyo, Tokyo

Hüsken D, Steudle E, Zimmermann U (1978) Pressure probe technique for measuring water relations of cells in higher-plants. Plant Physiol 61(2):158–163. https://doi.org/10.1104/pp.61.2.158

IPCC (2007) Climate change 2007: the physical science basis. In: Contribution of working group I to the fourth assessment report of the intergovernmental panel on climate change. Cambridge University Press, Cambridge/New York

Ishihara K, Horiguchi T, Mizuno S, Takahashi H (2005) On the occurence of milky white like kernels of rice plants under high temperature conditions with reference to water stress and nitrogen fertilizer application in 2003. Proc Jpn J Crop Sci 74:122–123

Ishimaru T, Hirabayashi H, Sasaki K, Ye C, Kobayashi A (2016) Breeding efforts to mitigate damage by heat stress to spikelet sterility and grain quality. Plant Prod Sci 19(1):12–21

Jagadish SVK, Craufurd PQ, Wheeler TR (2007) High temperature stress and spikelet fertility in rice (*Oryza sativa* L.). J Exp Bot 58(7):1627–1635

Jagadish SVK, Craufurd PQ, Wheeler TR (2008) Phenotyping parents of mapping populations of rice for heat tolerance during anthesis. Crop Sci 48(3):1140–1146

Jagadish SVK, Muthurajan R, Oane R, Wheeler TR, Heuer S, Bennett J, Craufurd PQ (2010) Physiological and proteomic approaches to address heat tolerance during anthesis in rice (*Oryza sativa* L.). J Exp Bot 61(1):143–156

Jagadish SVK, Murty MVR, Quick WP (2015) Rice responses to rising temperatures-challenges, perspectives and future directions. Plant Cell Envirn 38(9):1686–1698

Kobayashi A, Sonoda J, Sugimoto K, Kondo M, Iwasawa N, Hayashi T, Tomita K, Yano M, Shimizu T (2013) Detection and verification of QTLs associated with heat-induced quality decline of rice (*Oryza sativa* L.) using recombinant inbred lines and near-isogenic lines. Breed Sci 63(3):339–346

Li Y, Fan C, Xing Y, Yun P, Luo L, Yan B, Peng B, Xie W, Wang G, Li X, Xiao J, Xu C, He Y (2014) *Chalk5* encodes a vacuolar H^+-translocating pyrophosphatase influencing grain chalkiness in rice. Nat Genet 46(4):398–404. https://doi.org/10.1038/ng.2923

Li X, Lawas LMF, Malo R, Glaubitz U, Erban A, Mauleon R, Heuer S, Zuther E, Kopka J, Hincha DK, Jagadish KSV (2015) Metabolic and transcriptomic signatures of rice floral organs reveal sugar starvation as a factor in reproductive failure under heat and drought stress. Plant Cell Environ 38(10):2171–2192

MAFF (2018) Standard evaluation for rice varieties. The Ministry of Agriculture, Forestry and Fisheries, Japan, 2018

Maruyama A, Weerakoon WMW, Wakiyama Y, Ohba K (2013) Effects of increasing temperatures on spikelet fertility in different rice cultivars based on temperature gradient chamber experiments. J Agron Crop Sci 199(6):416–423

Matsui T, Omasa K (2002) Rice (*Oryza sativa* L.) cultivars tolerant to high temperature at flowering: anther characteristics. Ann Bot 89(6):683–687

Matsui T, Namuco OS, Ziska LH, Horie T (1997a) Effects of high temperature and CO_2 concentration on spikelet sterility in indica rice. Field Crop Res 51(3):213–219. https://doi.org/10.1016/S0378-4290(96)03451-X

Matsui T, Omasa K, Horie T (1997b) High temperature-induced spikelet sterility of japonica rice at flowering in relation to air temperature, humidity and wind velocity conditions. Jpn J Crop Sci 66(3):449–455. https://doi.org/10.1626/jcs.66.449

Matsui T, Omasa K, Horie T (2001) The difference in sterility due to high temperatures during the flowering period among japonica-rice varieties. Plant Prod Sci 4(2):90–93. https://doi.org/10.1626/pps.4.90

Morita S, Nakano H (2011) Nonstructural carbohydrate content in the stem at full heading contributes to high performance of ripening in heat-tolerant rice cultivar Nikomaru. Crop Sci 51(2):818–828. https://doi.org/10.2135/cropsci2010.06.0373

Morita S, Shiratsuchi H, Takahashi J, Fujita K (2004) Effect of high temperature on grain ripening in rice plants: analysis of the effects of high night and high day temperatures applied to the panicle and other parts of the plant. Jpn J Crop Sci 73(1):77–83. https://doi.org/10.1626/jcs.73.77

Morita S, Yonemaru J, Takanashi J (2005) Grain growth and endosperm cell size under high night temperatures in rice (*Oryza sativa* L.). Ann Bot 95(4):695–701. https://doi.org/10.1093/aob/mci071

Morita S, Wada H, Matsue Y (2016) Countermeasures for heat damage in rice grain quality under climate change. Plant Prod Sci 19(1):1–11

Nagato K, Chaudhry FM (1970) Influence of panicle clipping, flag leaf cutting and shading on ripening of Japonica and Indica rice. Proc Crop Sci Soc Jpn 39:204–212

Nakagawa H, Horie T, Matsui T (2003) Effects of climate change on rice production and adaptive technologies. In: Mew TW, Brar DS, Peng S, Dawe D, Hardy B (eds) Rice science: innovations and impact for livelihood. Proceedings of international rice research conference. International Rice Research Institute, Los Baños, pp 635–658

Nakashima T, Wada H, Morita S, Erra-Balsells R, Hiraoka K, Nonami H (2016) Single-cell metabolite profiling of stalk and glandular cells of intact trichomes with internal electrode capillary pressure probe electrospray ionization mass spectrometry. Anal Chem 88(6):3049–3057

Nishiyama I, Satake T (1981) High-temperature damages in rice plants. Jpn J Trop Agr 25(1):14–19

Nonami H, Boyer JS (1993) Direct demonstration of a growth-induced water potential gradient. Plant Physiol 102(1):13–19

Prasad PVV, Boote KJ, Allen LH, Sheehy JE, Thomas JMG (2006) Species, ecotype and cultivar differences in spikelet fertility and harvest index of rice in response to high temperature stress. Field Crop Res 95(2–3):398–411

Sakai M, Okamoto M, Tamura K, Kaji R, Mizobuchi R, Hirabayashi H, Fukaura S, Nishimura M, Yagi T (2007) "Nikomaru", a new rice variety with excellent palatability and grain appearance developed for warm region of Japan. Breed Res 9:67–73

Satake T, Yoshida S (1978) High temperature-induced sterility in indica rices at flowering. Jpn J Crop Sci 47(1):6–17. https://doi.org/10.1626/jcs.47.6

Sato H, Ishii T, Ohta H, Maeda H, Ideta O, Takeuchi Y, Kaji R, Nagaoka I, Hirabayashi H, Shigemune A, Tamura K (2018) Selection of rice standard varieties for evaluating grain quality under high temperature, employed in the examination of applied variety of MAFF. In: Proceedings of Japanese society of breeding. Fukuoka, Japan, March 2018, pp 223, P099

Sreenivasulu N, Butardo VM, Misra G, Cuevas RP, Anacleto R, Kishor PBK (2015) Designing climate-resilient rice with ideal grain quality suited for high-temperature stress. J Exp Bot 66(7):1737–1748

Steudle E (1993) Pressure probe techniques: basic principles and application to studies of water and solute relations at the cell, tissue and organ level. In: Smith J, Griffiths H (eds) Water deficits: plant responses from cell to community. Bios Scientific Publishers, Oxford, pp 5–36

Tashiro T, Wardlaw IF (1991) The effect of high-temperature on kernel dimensions and the type and occurrence of kernel damage in rice. Aust J Agric Res 42(3):485–496

Tomos AD, Leigh RA (1999) The pressure probe: a versatile tool in plant cell physiology. Annu. Rev. Plant Physiol. Plant Mol. Biol. 50(1):447–472

Usui Y, Sakai H, Tokida T, Nakamura H, Nakagawa H, Hasegawa T (2014) Heat-tolerant rice cultivars retain grain appearance quality under free-air CO_2 enrichment. Rice 7:6

Wada H, Nonami H, Yabuoshi Y, Maruyama A, Tanaka A, Wakamatsu K, Sumi T, Wakiyama Y, Ohuchida M, Morita S (2011) Increased ring-shaped chalkiness and osmotic adjustment when growing rice grains under foehn-induced dry wind condition. Crop Sci 51(4):1703–1715. https://doi.org/10.2135/cropsci2010.08.0503

Wada H, Masumoto-Kubo C, Gholipour Y, Nonami H, Tanaka F, Erra-Balsells R, Tsutsumi K, Hiraoka K, Morita S (2014) Rice chalky ring formation caused by temporal reduction in starch biosynthesis during osmotic adjustment under foehn-induced dry wind. PLoS One 9(10):e110374. https://doi.org/10.1371/journal.pone.0110374

Wada T, Inoue T, Tsubone M, Ogata T, Miyahara K, Hamachi Y, Furusho M, Miyazaki M, Yamaguchi O, Ishibashi M, Sato H, Matsue Y (2016) 'Minoritsukushi', a new rice cultivar with medium-late maturity, high yield, fine palatability and tolerance to high temperature during ripening period. Bull Fukuoka Agric For Res Center 2:1–7

Wada H, Hatakeyama Y, Onda Y, Nonami H, Nakashima T, Erra-Balsells R, Morita S, Hiraoka K, Tanaka F, Nakano H (2019) Multiple strategies for heat adaptation to prevent chalkiness in the rice endosperm. J Exp Bot 70(4):1299–1311

Wakamatsu K, Sasaki O, Uezono I, Tanaka A (2008) Effect of the amount of nitrogen application on occurrence of white-back kernels during ripening of rice under high-temperature conditions. Jpn J Crop Sci 77(4):424–433. https://doi.org/10.1626/jcs.77.424

Wakamatsu K, Yamane I, Sato M, Komaki Y, Ohuchida M, Mori K, Sonoda J, Goto H, Shigemizu T, Kuwahara H, Tanaka A, Nagayoshi S (2016) Breeding a new rice cultivar 'Natsuhonoka'. Bull Kagoshima Pref Ins Agric Dev 10:9–20

Watson A, Ghosh S, Williams MJ, Cuddy WS, Simmonds J, Rey M-D, Asyraf Md Hatta M, Hinchliffe A, Steed A, Reynolds D, Adamski NM, Breakspear A, Korolev A, Rayner T, Dixon LE, Riaz A, Martin W, Ryan M, Edwards D, Batley J, Raman H, Carter J, Rogers C, Domoney C, Moore G, Harwood W, Nicholson P, Dieters MJ, DeLacy IH, Zhou J, Uauy C, Boden SA, Park RF, Wulff BBH, Hickey LT (2018) Speed breeding is a powerful tool to accelerate crop research and breeding. Nat Plants 4(1):23–29. https://doi.org/10.1038/s41477-017-0083-8

Weerakoon WMW, Maruyama A, Ohba K (2008) Impact of humidity on temperature-induced grain sterility in rice (*Oryza sativa* L). J Agron Crop Sci 194(2):135–140

Yang YY, Huang YY, Wu JH, Liu N, Deng JW, Luan TG (2017) Single-cell analysis by ambient mass spectrometry. Trac-Trend Anal Chem 90:14–26

Chapter 4
Controlling the Depth of Soil Frost in Farm Fields in Japan

Tomotsugu Yazaki and Tomoyoshi Hirota

Abstract Winter climate change has resulted in earlier onset of deep snow cover and reduced soil-frost depth in eastern Hokkaido, Japan, since the late 1980s. In the Tokachi region, a major potato-producing area of Japan where upland fields are managed by rotation, small potato tubers unharvested in the fall survive the winter and emerge as weeds (volunteer potatoes) during the next crop season and cause problems for producers. This is attributable to the thermal insulation of soil by the thick snowpack and inhibits unharvested potato tubers from freezing to die. To develop an adaptive countermeasure for these problems, a soil-frost depth control technique, which manipulates snow cover thickness and controls soil-frost depth to eliminate volunteer potatoes, was developed. Field trials demonstrated that soil-frost depths can be predicted with an accuracy of several centimeters. A target soil-frost depth of 0.30 m is proposed for complete elimination of volunteer potatoes. A numerical model facilitates decision-making related to the scheduling of snow-plowing practices for freeze death of potato tubers. This method has been adopted by local potato producers who manage farmland on a large scale. It represents a new agricultural technology that is useful for adaptation to climate change.

4.1 Introduction

The changes in climate during winter affect the productivity of natural vegetation and agricultural crops in cold regions (e.g., Imai et al. 2013). In high-latitude and cold regions, where crop species grow below their optimal temperature, crop

T. Yazaki (✉)
School of Agriculture, Meiji University, Kawasaki, Kanagawa, Japan
e-mail: yazakitm@meiji.ac.jp

T. Hirota
NARO Hokkaido Agricultural Research Center, Sapporo, Japan

T. Iizumi et al. (eds.), *Adaptation to Climate Change in Agriculture*,
https://doi.org/10.1007/978-981-13-9235-1_4

productivity can increase as rising temperatures reduce the probability of low temperature, a factor reducing yield (Murray and Gaudet 2013). However, climate change often indirectly affects agricultural activities, such as pests and weeds. In the case of the Tokachi region in northern Japan, soil-frost depth has been decreasing since the late 1980s (Fig. 4.1, Hirota et al. 2006), facilitating the overwinter survival of small potato (*Solanum tuberosum* L.) tubers that remain unharvested. They sprout and emerge from the soil surface as weeds, called volunteer potatoes, during the subsequent cropping season (Fig. 4.2, Hirota et al. 2011; Hirota et al. 2013). This is attributable to the thermal insulation of soil by the thick snowpack with low thermal conductivity and keeps soil temperature higher. Therefore, to lower the soil temperature, removing snowpack and enhancing heat loss from the soil are necessary. To eliminate such volunteer potatoes, snow plowing (Fig. 4.3, hereinafter, *yukiwari* in Japanese), which removes snow cover and allows deeper soil frost, is spreading as a practical countermeasure in the Tokachi region (Hirota et al. 2011, 2013;

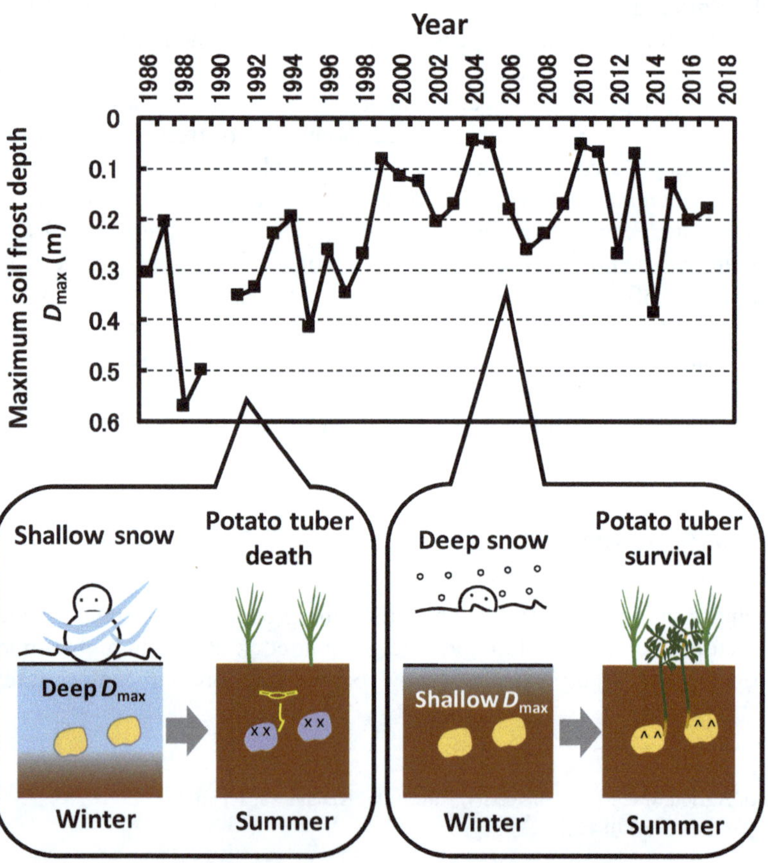

Fig. 4.1 Annual maximum soil-frost depth (D_{max}) at Hokkaido Agricultural Research Center in Memuro, in the Tokachi region of Hokkaido, Japan

Fig. 4.2 Photographs showing (a) potatoes left unharvested in the fall, and (b) volunteer potatoes emerging in a sugar beet field, (c) in a wheat field, and (d) in a bean field. (Hirota et al. 2011)

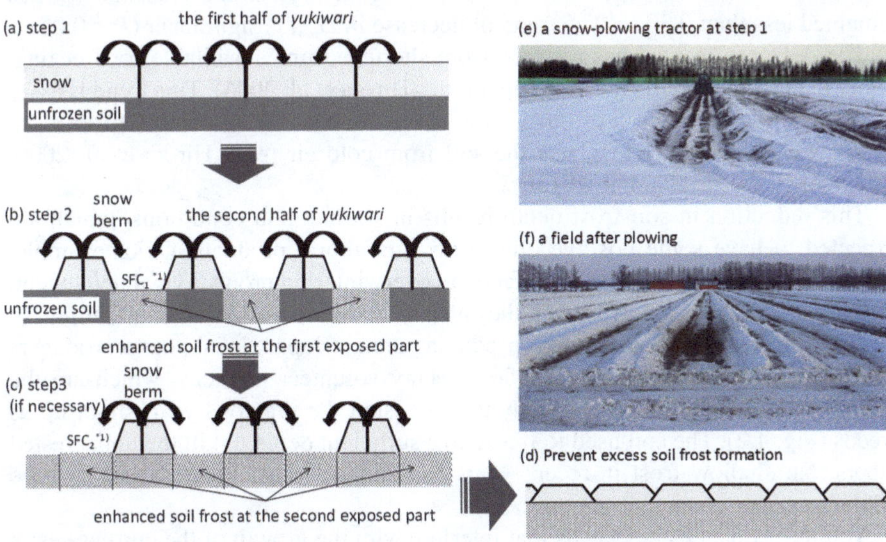

Fig. 4.3 Schematic diagrams of soil-frost control showing (**a**)–(**d**) the operation sequence of *yukiwari*, and the photographs showing (**e**) a snow-removing tractor and (**f**) the field after *yukiwari*. (Modified from Yazaki et al. 2013a and Hirota et al. 2013)

*1) SFC_1 and SFC_2 are exposed soil surfaces after the first and second steps

Yazaki et al. 2013a; Shirahata 2018). In earlier stages, however, the technique was empirical, and the level of expertise varies among potato producers. The unharvested potato tubers can survive the winter if the soil frost is too shallow (Yazaki et al. 2013a). On the contrary, if the soil frost is too deep, it can delay spring agricultural practices and the onset of crop growth. To avoid these negative effects, we developed a soil-frost control method by manipulating snow cover thickness using a numerical soil temperature model. In this chapter, we present (1) the reduction in soil-frost depths and volunteer potato problems, (2) the method of soil-frost control that attained the target soil-frost depth, (3) an assessment of the work schedule for the final limit of the work based on meteorological data and models, and (4) a method of enhancing soil-frost control technique by linking it to agrometeorological information.

4.2 Reducing Soil-Frost Depths and Volunteer Potato Problems

The winter climate in Japan has changed greatly since the late 1980s. For example, in the Tokachi region of northern Japan, the soil-frost depth has remarkably decreased since the 1990s. The annual maximum soil-frost depth (D_{max}) at the NARO Hokkaido Agricultural Research Center in Memuro, in the central part of the Tokachi region, shows that the D_{max} has drastically decreased (Fig. 4.1). The D_{max} in the 1980s was generally greater than 0.30 m. From 1998 to the present, D_{max} has remained less than 0.30 m. The trend of decrease in D_{max} is significant ($P < 0.05$).

The decrease in D_{max} has resulted from shortened times available for freezing, rather than from increases in air temperature (Hirota et al. 2006). This trend is associated with the early onset of deep snow cover of about 0.2 m depth (Hirota et al. 2006), which thermally insulates the soil from cold air (e.g., Hirota et al. 2006; Tsuchiya 1985; Yazaki et al. 2013b).

This reduction in soil-frost depth results in warmer soil conditions, which are expected to have some positive effects on agricultural production. For example, warmer conditions have positive effects on overwintering crop survival and extend the crop growing season. However, they also have negative effects on agriculture. In the Tokachi region, a major potato-producing area of Japan, local potato producers have been facing struggles in the field. Many volunteer potatoes, which are the sprouts from small tubers that were not harvested the previous year, shoot up as weeds (Fig. 4.2). The soil used to freeze to a sufficient depth to kill the unharvested tubers, but shallow frost in recent years has allowed some tubers to survive the winter.

Volunteer potatoes are weeds that interfere with the growth of the current year's rotation crops. They can cause pest problems by attracting aphids, transferring viruses, and hosting potato cyst nematodes. They might also become a risk to trade, because they can cause undesired mixing when different breeds of potato are planted in consecutive years. Therefore, it is necessary to completely remove volunteer potatoes. Local producers have been forced to engage in manual weeding, which requires tens of hours per hectare. Because local farmers grow potatoes on an area larger than several hectares, on average, the manual removal of volunteer potatoes necessitates a severe increase in workload during the busy farming period.

4.3 Theory and Effects of Soil-Frost Control for Volunteer Potato Weeding

4.3.1 Determination of Target Soil-Frost Depth

Some local potato producers in the Tokachi region adopted countermeasures to address the decreased soil-frost depth in the 1990s (Shirahata 2018). They started plowing snow, which temporarily removes snow cover to allow the frost to penetrate deeper and consequently kill volunteer potato tubers. Producers can enhance soil-frost depth even in large fields using machines such as plow-mounted tractors or bulldozers. Although the method appears to be a breakthrough supporting the confrontation of environmental change, this method of weeding is not stably effective because it was based on trial and error and relied on the producers' experience and intuition. The potato tubers survive the winter if the soil frost is too shallow. In contrast, if the soil frost is too deep, it delays spring agricultural practices and crop growth.

To stabilize the effects of volunteer potato control and prevent adverse effects of prolonged soil freezing, it is necessary to develop techniques based on scientific evidence, such as target and estimated soil-frost depth.

A numerical soil temperature model (extended force-restore model) has been developed (Hirota et al. 2002). Using the soil temperature model, the soil temperature and soil-frost depth can be estimated using air temperature and snow cover thickness as input data. The soil temperature model enabled us to calculate the best day for *yukiwari* to achieve the target soil-frost depth by figuring out the soil temperature profile and soil-frost depth.

To determine the optimized and target soil-frost depths, it is crucial to obtain the unharvested tuber distribution and critical temperature necessary to kill potato tubers. Most unharvested tubers, 97% and 99.9% on average at the four sites in fall 2010 and 2011, were located in a layer at a depth of less than 0.15 m from the soil surface (Yazaki et al. 2013a). We thereby confirmed that most tubers were distributed at depths of 0.15 m or less. Some reports have suggested that the critical temperature must be below −3 °C to kill potato tubers (Li et al. 1981; Boydston et al. 2006; Hirota et al. 2011). Accordingly, the critical soil temperature was set to −3 °C.

By combining the critical temperature and the estimated soil temperature profile, a new indicator of the critical frost depth to kill volunteer potatoes was introduced (Hirota et al. 2011). Using the soil temperature model, the critical soil-frost depth, which is the soil-frost depth that would lead to the critical temperature of −3 °C at a depth of 0.15 m, was calculated to be 0.30 m. Therefore, to completely eradicate the volunteer potatoes, a maximum soil-frost depth of more than 0.30 m must be attained.

4.3.2 Procedure of Soil-Frost Control

The critical soil-frost depth provides information related to condition of volunteer potato weeding in fields. Based on field observations and analyses, an optimal soil-frost depth of 0.28–0.32 m has been proposed as a compromise between eradication of volunteer potatoes and minimizing negative effects on agriculture (Yanai et al. 2017), such as delay of agricultural practices in the spring, N leaching and pollution in groundwater (e.g., Iwata et al. 2013, 2018), and nitrous oxide emissions (Yanai et al. 2011). Here, we set the target soil-frost depth as 0.30 m. If the target soil-frost depth is attained by soil-frost control, almost all tubers can be eliminated, while minimizing the negative effects of freezing.

The procedure for soil-frost control based on the simulated soil-frost depth (SFD_m) is shown in Fig. 4.3 (Yazaki et al. 2013a). The first half of the *yukiwari* (snow plowing in a line, leaving snow berms on both sides; Fig. 4.3a) step is started when the snow cover thickness exceeded 0.20 m, the thickness above which additional soil-frost formation is inhibited (e.g., Hirota et al. 2006; Tsuchiya 1985). Then, we leave the first exposed part (hereinafter, SFC_1) uncovered by snow until the SFD_m reach the target depth; SFD_m is confirmed using a soil temperature model. In cases where the SFC_1 is covered with fresh snow deeper than 0.10 m, the snow is plowed to prevent inhibition of rapid soil-frost penetration. When the SFD_m at SFC_1 reaches 0.30 m, we move on to the second step (Fig. 4.3b). The snow berms formed in the first step are plowed to expose the soil surface (the second exposed part: hereinafter SFC_2; Fig. 4.3b). The soil surface at SFC_2 is exposed until the SFD_m reaches the target depth. When SFC_2 is covered with fresh snow deeper than 0.10 m, the snow is plowed to deepen soil frost, as it is for SFC_1. Therefore, work is finished when SFD_m reached the target depth of 0.30 m at both SFC_1 and SFC_2 (Fig. 4.3c). Excess deep penetration of the freezing front sometimes induces negative effects on farming, such as a delay in seeding after snowmelt (Hirota et al. 2011), and enhances soil erosion by the surface movement of snowmelt water (Iwata et al. 2010). If excess soil-frost formation is observed after reaching the target depth, the snow berm was plowed to cover the second exposed part to prevent additional soil-frost formation (Fig. 4.3c, d).

4.3.3 Field Experiment of Soil-Frost Control

A field experiment was carried out at four sites for 2 years in the Tokachi region of Japan. These fields were previously under potato cultivation. The number of unharvested potato tubers was counted in late November of 2010 and 2011. The air temperature and snow cover thickness were measured at each station from late November to April during the winters of 2010/11 and 2011/12. The field was divided into two subplots and allocated to a soil-frost control plot (SFC) and a snow-undisturbed plot (control plot: CO). In the SFC plot, soil-frost control was carried out to accomplish the target soil-frost depth of 0.30 m, guided by the soil-frost depth calculated using the soil temperature model (SFD_m). The actual soil-frost depths (SFD_a) at the first and second exposed parts were manually measured by soil sampling with auger. In the next cropping season, the number of sprouting volunteer potatoes was counted two to four times in June–July in both the CO plot and SFC plot at each site.

Sequential changes in SFD_m and SFD_a agreed fairly well (Fig. 4.4). Furthermore, maximum SFD_m coincided well with that of SFD_a with an RMSE of 0.067 m. The maximum SFD_m and SFD_a in the CO plots were less than the target depth of 0.30 m, except at Shihoro in the winter of 2010–2011, which had less than 0.20 m snow cover thickness. Therefore, *yukiwari* was not carried out. In contrast, the maximum SFD_m and SFD_a in the SFC plot exceeded 0.30 m after snow removal was practiced 1–6 times in the SFC_1 plots and 1–2 times in the SFC_2 plots, in most cases.

The number of sprouts that emerged during the following cropping season in the CO plot (0–2.86 tubers m^{-2}) was much lower than that of unharvested leftover tubers (4.73–18.45 tubers m^{-2}). In the SFC plot, the number of sprouts that emerged during the following cropping season was much smaller than that at the CO plot. No volunteer potato sprouts emerged in either the CO plot or SFC plot at Shihoro in

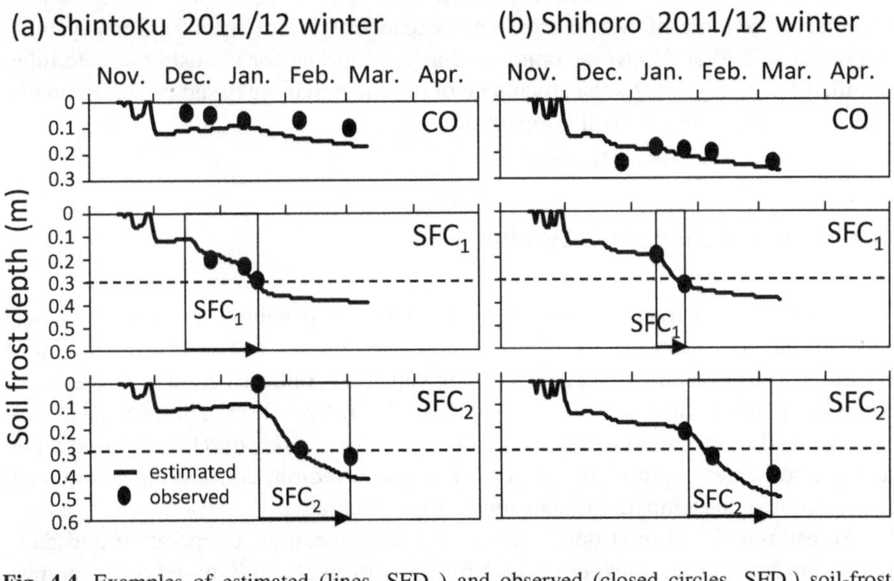

Fig. 4.4 Examples of estimated (lines, SFD_m) and observed (closed circles, SFD_a) soil-frost depths in soil-frost control (SFC_1 and SFC_2) and undisturbed control (CO) plots. Data are from the (**a**) Shintoku and (**b**) Shihoro site in the winter of 2011/12. Broken lines in the SFC_1 and SFC_2 panels denote the target soil-frost depth of 0.30 m. Rectangles with arrows in each panel show the time period during which soil surface was exposed for accelerated soil-frost formation. (Redrawn based on Yazaki et al. 2013a)

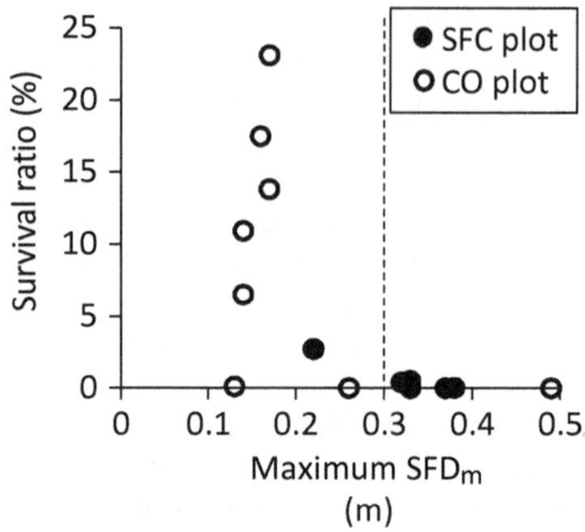

Fig. 4.5 Relationship between maximum soil-frost depth (SFD_m) and survival ratio of unharvested potato tubers for all study sites over 2 years of observation. Closed and open circles, respectively, denote values for the SFC plot (SFC_1 or SFC_2, whichever is smaller) and the CO plot. The dashed line shows the target SFD_m of 0.30 m. (Redrawn based on Yazaki et al. 2013a)

either year. The survival ratio, which was the ratio of the number of emerged sprouts to that of unharvested tubers, was 0–2.7% in the SFC plot, which was significantly smaller ($P < 0.01$) than that in the CO plots (2.4–23.1%), with the exception of Shihoro. The survival ratio was almost zero when SFD_m exceeded 0.30 m.

Soil-frost control remarkably decreased survival to less than 0.5% of the number of unharvested tubers if the soil-frost depth exceeded 0.30 m (Fig. 4.5). Consequently, the target soil-frost depth was confirmed as appropriate for almost complete tuber elimination. This suggests the advantage of decision-making based on the estimated soil-frost depth using the soil temperature model.

4.4 Work Schedule Guidelines

To diffuse the soil-frost control technique of weeding volunteer potatoes to local potato producers, failure in weeding caused by shortage of soil-frost formation must be minimized. To attain stable effects using soil-frost control, a guideline is required to create a statistically safe schedule for *yukiwari* operation. Therefore, we proposed a final limit for the second half of the process of *yukiwari* (Fig. 4.3b) to attain a target soil-frost depth of 0.3 m for the Tokachi region, considering interannual fluctuations in air temperature and snow cover thickness.

We estimated soil-frost depths using the daily mean air temperature and daily maximum snow cover thickness from November through April, based on the numerical soil temperature model (Hirota et al. 2002). These input data were obtained from

18 meteorological stations over 23 years in Tokachi (JMA 2012). The final limit date of the second half-phase of *yukiwari* to expose SFC_2 (Fig. 4.3b) in each year was determined from the final date on which the soil-frost depth reached 0.3 m, assuming that the snow cover thickness after *yukiwari* was 0.02 m. For example, in the case of Memuro in winter 2011, *yukiwari* work on February 6 attained 0.3 m soil-frost depth, but work on February 7 did not. In that case, the final limit date in the 2011 winter was determined as February 6 (data not shown). To obtain the final limit for *yukiwari* even under extremely warm winter conditions, the statistically determined final limit day with a return period of 30 years was calculated by application of the final limit dates for 23 years to a Weibull distribution (Yamada et al. 1998).

In the Tokachi region, the soil-frost depth reached 0.30 m using *yukiwari* in most areas except Hiroo, in the southernmost coastal area (Fig. 4.6). The final limit in a 30-year return period was estimated to be around February 20 in mountainous areas in the north and from late January to early February in other areas (Fig. 4.6). The use of soil temperature model is useful in scheduling *yukiwari* to minimize failure of soil-frost control for weeding volunteer potatoes.

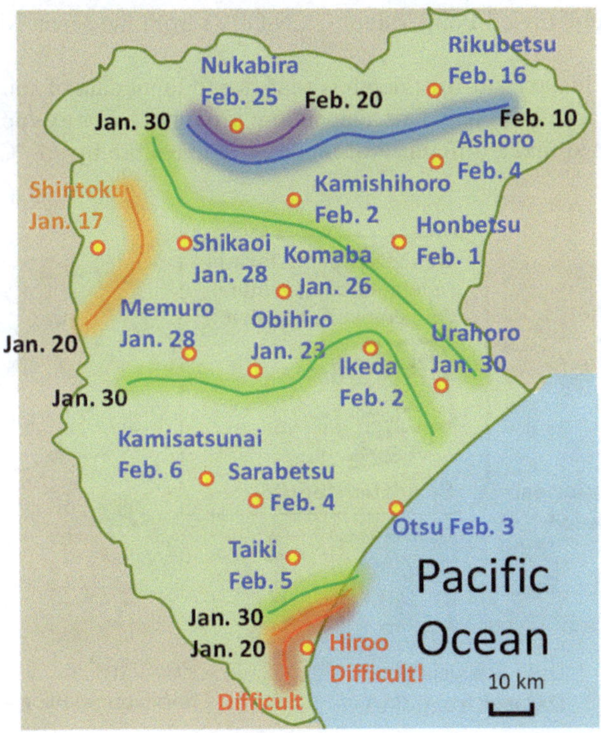

Fig. 4.6 Final limit of *yukiwari* in a return period of 30 years for the Tokachi region. (Redrawn based on Yazaki et al. 2012)

4.5 Enhancing Soil-Frost Control Technique: Linkage with the Agrometeorological Grid Square Data System and Improvements in Accuracy

Recent information technology (IT) development and diffusion enable access to agrometeorological information by local producers with smartphones or other devices to aid in decision-making in their work (Hansen and Coffey 2011; Kobayashi et al. 2010). In soil-frost control for the purpose of volunteer potato weeding, local producers can access information on soil-frost depths in their own fields from the web server of the Tokachi Federation of Agricultural Cooperative Association (Fig. 4.7). To accurately predict soil-frost depth, accurate air temperature, and snow cover thickness are necessary to obtain the input data for the soil temperature model. Especially, air temperature is sensitive to the simulated soil temperature and soil-frost depth.

Agrometeorological information has been developed and utilized in various situations, such as predicting plant growth or predicting crop disease (e.g., Breuer et al. 2008). Recently, the Agro-Meteorological Grid Square Data system was developed (Ohno et al. 2016). Also, the 1 km gridded air temperature data for the normal and the predicted daily air temperatures for up to 9 days ahead were provided by Global Spectral Model-Gridded Point Value (GMS-GPV) from the Japan Meteorological Agency.

Because estimated soil-frost depth is sensitive to input data of soil temperature, the high accuracy of estimated air temperature is required. For example, sensitivity analysis revealed that a difference of input air temperature by ±1 °C produces an

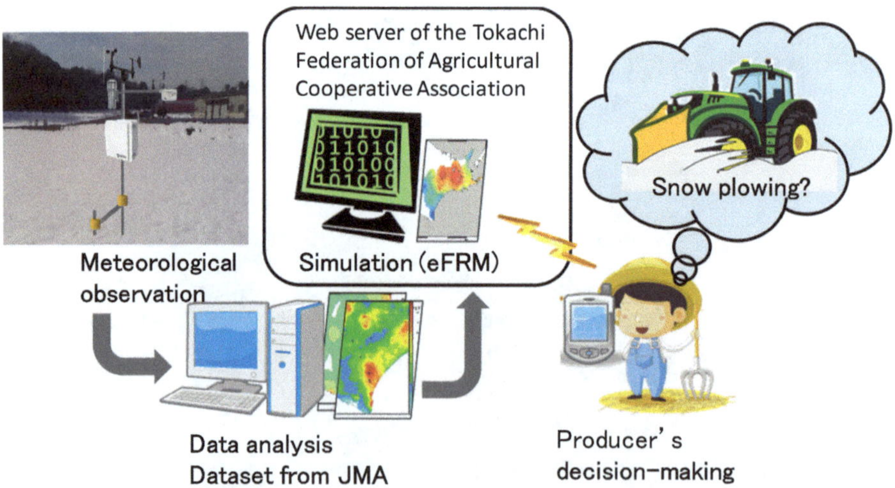

Fig. 4.7 Schematic figure of soil-frost control by utilization of agro-information for producers' decision-making. This system is utilized by local farmers to guide soil-frost depth control

error of 2–4 cm in soil-frost depth (Hirota et al. 2011). It has been reported that the air temperature in the Tokachi region varies locally, even though there is little difference in altitude (Miwa 1963; Sorai et al. 2016; Yazaki et al. 2017; Fukushima et al. 2019). Meteorological observations have revealed that the difference in daily mean air temperatures often exceeds 5 °C between two meteorological stations separated by a distance of ca. 5 km along a gentle slope (Yazaki et al. 2011). Data from 19 JMA meteorological stations showed that the winter mean of the daily mean air temperature was strongly correlated with the winter mean of the daily minimum air temperature. The means of the daily minimum air temperatures were not correlated with altitude or distance from the sea. This suggests that locally varying daily minimum air temperatures strongly influence the winter mean air temperature in the Tokachi region.

As shown above, it is required to estimate accurate air temperature distribution to calculate soil-frost depth at the Tokachi region where spatial variation of air temperature is large. Improvements of estimation of air temperature distribution would enhance producer's decision-making system based on the soil temperature model.

4.6 Conclusions

To address the problem of increased volunteer potato growth caused by decreased soil-frost depth, the complete eradication of volunteer potatoes has become possible by using soil-frost control based on a scientifically obtained target soil-frost depth and soil temperature simulation. This practice not only controls volunteer potatoes, but also greatly reduces the workload of local producers because it is practiced during winter, an agriculturally slack season.

Furthermore, presenting the final limit date for the second half of *yukiwari* is useful in helping local potato producers to confidently introduce the practice.

Soil-frost control for weeding volunteer potatoes has diffused to an estimated 60–70% of the area (approximately 5000 hectares) where it can be implemented, mainly in the Tokachi region. Using the Internet to communicate, local producers in their field can now determine soil temperature and soil-frost depth by accessing soil temperature estimation models on the web server. For appropriate use of agrometeorological information, simulation or estimation techniques for air temperature are required for the Tokachi plain, where local variation in temperature is large.

Acknowledgments We are grateful for the support received from Y. Iwata, S. Inoue, M. Shirahata, T. Suzuki, K. Araki, T. Kajiyama, K. Maezuka, and I. Yamanaka. We also acknowledge the valuable advice received from S. Hasegawa and N. Matsumoto. We thank the editor and two anonymous reviewers for their constrictive comments. This research was supported by grants from the Research and Development Project for Application in Promoting New Policy for Agriculture, Forestry, and Fisheries (22079) from the Japanese Ministry of Agriculture, Forestry and Fisheries, and a project by the NARO Bio-oriented Technology Research Advancement Institution (research program on development of innovative technology (29017C)).

References

Boydston RA, Seymour MD, Brown CR et al (2006) Freezing behavior of potato (Solanum tuberosum) tubers in soil. Am J Potato Res 83:306–315

Breuer NE, Cabrera VE, Ingram KT et al (2008) AgClimate: a case study in participatory decision support system development. Clim Chang 87:385–403

Fukushima H, Yazaki T, Hirota T et al (2019) Factors and mechanisms affecting the air temperature distribution on a clear winter night in a snow-covered mesoscale plain. J Meteorol Soc Jpn Ser II 97:105–121. https://doi.org/10.2151/jmsj.2019-005

Hansen J, Coffey K (2011) Agro-climate tools for a new climate-smart agriculture. https://ccafs.cgiar.org/sites/default/files/assets/docs/iri_agro-climate_tools_brief-web.pdf. Accessed 20 Nov 2018

Hirota T, Pomeroy JW, Granger RJ et al (2002) An extension of the force-restore method to estimating soil temperature at depth and evaluation for frozen soils under snow. J Geophys Res 107(D24):4767. https://doi.org/10.1029/2001JD001280

Hirota T, Iwata Y, Hayashi M et al (2006) Decreasing soil-frost depth and its relation to climate change in Tokachi, Hokkai-do, Japan. J Meteorol Soc Jpn 84:821–833

Hirota T, Usuki K, Hayashi M et al (2011) Soil frost control: agricultural adaptation to climate variability in a cold region of Japan. Mitig Adapt Strateg Glob Chang 16:791–802

Hirota T, Yazaki T, Usuki K et al (2013) Soil frost control: its application to volunteer potato management in a cold region. In: Imai R, Yoshida M, Matsumoto N (eds) Plant and microbe adaptations to cold in a changing world: proceedings of the plant and microbe adaptation to cold conference. Springer, New York, pp 51–60

Imai R, Yoshida M, Matsumoto N (eds) (2013) Plant and microbe adaptations to cold in changing world. Springer, New York

Iwata Y, Hayashi M, Suzuki S et al (2010) Effects of snowcover on soil freezing, water movement and snowmelt infiltration: a paired plot experiment. Water Resour Res 46:W09504. https://doi.org/10.1029/2009WR008070

Iwata Y, Yazaki T, Suzuki S et al (2013) Water and nitrate movements in an agricultural field with different soil frost depths: field experiments and numerical simulation. Ann Glaciol 54:157–165

Iwata Y, Yanai Y, Yazaki T et al (2018) Effects of a snow-compaction treatment on soil freezing, snowmelt runoff, and soil nitrate movement: a field-scale paired-plot experiment. J Hydrol 567:280–289

Japanese Meteorological Agency (JMA) (2012) Source of the archived meteorological data. http://www.data.jma.go.jp/obd/stats/etrn/index.php. Accessed 4 Jan 2012

Kobayashi T, Kanno H, Kanda E et al (2010) Development of rice climate damage reduction system using weather forecasting data based on Google Maps. Plant Prot 64:794–800 (in Japanese)

Li PH, Huner NPA, Toivio-Kinnucan M et al (1981) Potato freezing injury and survival, and their relationship to other stress. Am J Potato Res 58:15–29

Miwa K (1963) Winter temperature in Tokachi, Hokkaido. Tenki 10:381–384 (in Japanese)

Murray T, Gaudet D (2013) Global change in winter climate and agricultural sustainability. In: Imai R, Yoshida M, Matsumoto N (eds) Plant and microbe adaptations to cold in a changing world: proceedings of the plant and microbe adaptation to cold conference. Springer, New York, pp 1–15

Ohno H, Sasaki K, Ohara G et al (2016) Development of grid square air temperature and precipitation data compiled from observed, forecasted and climatic normal data. Clim Biosph 16:71–79. in Japanese with English abstract

Shirahata M (2018) Review of volunteer potatoes. Hokkaido J Agric Meteorol 70:12–21 (in Japanese)

Sorai T, Hamada H, Kameda T et al (2016) Rikubetsu, the coldest town in Japan: according to the meteorological data by Japan meteorological agency from 2007 to 2016 in winter seasons. Tenki 63:879–887 (in Japanese)

Tsuchiya F (1985) Effect of soil frost and frost penetration on volcanic ash soil in agricultural field in Tokachi, Eastern Hokkaido, Japan. Dissertation, Hokkaido University (in Japanese)

Yamada K, Hirota T, Shirooka T (1998) Analyses of continuous snow cover duration and its first and last days using Weibull distribution method. J Agric Meteorol 54:63–69 (in Japanese with English abstract)

Yanai Y, Hirota T, Iwata Y et al (2011) Accumulation of nitrous oxide and depletion of oxygen in seasonally frozen soils in northern Japan: snow cover manipulation experiments. Soil Biol Biochem 43:1779–1786

Yanai Y, Iwata Y, Hirota T (2017) Optimum soil frost depth to alleviate climate change effects in cold region agriculture. Sci Rep 7:-44860

Yazaki T, Hirota T, Inoue S et al (2011) Relation among local nocturnal cooling, net radiation, and wind speed in the Tokachi region, Hokkaido. Proceedings of the annual meeting of the Hokkaido society of agricultural meteorology B1–B2 (in Japanese)

Yazaki T, Hirota T, Suzuki T et al (2012) Work schedule of yukiwari (snow plowing) to weed volunteer potatoes by soil-frost control using a numerical model. Clim Biosph 12:12–20 (in Japanese with English abstract)

Yazaki T, Hirota T, Iwata Y et al (2013a) Effective killing of volunteer potato (Solanum tuberosum L.) tubers by soil frost control using agrometeorological information - an adaptive countermeasure to climate change in a cold region. Agric For Meteorol 182–183:91–100

Yazaki T, Iwata Y, Hirota T et al (2013b) Influences of winter climatic conditions on the relation between annual mean soil and air temperatures from central to northern Japan. Cold Reg Sci Technol 85:217–224

Yazaki T, Fukushima H, Hirota T et al (2017) Winter nocturnal air temperature distribution for a mesoscale plain of a snow-covered region: field meteorological observations and numerical simulations. J Appl Meteorol Climatol 56:519–533

Chapter 5
Impact Assessment of Climate Change on Rice Yield Using a Crop Growth Model and Activities Toward Adaptation: Targeting Three Provinces in Indonesia

Yoshiyuki Kinose and Yuji Masutomi

Abstract In accordance with the Paris Agreement, the assessment of climate change impacts on rice productivity is expected at both the national and local (i.e., province or state) levels in Asian countries, to create and implement adaptation plans for climate change. However, there is limited information on the impact of climate change on local rice production in developing countries, especially at the local level. To this end, we aimed to clarify this impact on the yield of local rice in Indonesia in the next 25 years, using a crop growth model, MATCRO-Rice, in three provinces including North Sumatra, East Java, and Bali. Climate change was predicted to reduce the yield primarily because of increase in air temperature. Furthermore, the simulated yield reductions were different among the districts in each province, indicating the importance of regional adaptation priorities. We discussed several adaptation strategies with local stakeholders in each province from the viewpoints of feasibility and priority. Some strategies, such as change in cultivars to have high tolerance to high air temperature, which was ranked as being highly feasible and high priority, are expected to be future adaptation options.

5.1 Introduction

The concentration of atmospheric carbon dioxide (CO_2) has increased up to approximately 400 ppm owing to biomass burning and fossil fuel combustion (Joos and Spahni 2008; Dlugokencky and Tans 2018). Climate change such as the increase in air temperature and change in precipitation have occurred concurrently with the

Y. Kinose
Graduate Faculty of Interdisciplinary Research, University of Yamanashi, Kofu, Yamanashi, Japan

Y. Masutomi (✉)
College of Agriculture, Ibaraki University, Inashiki, Ibaraki, Japan

T. Iizumi et al. (eds.), *Adaptation to Climate Change in Agriculture*,
https://doi.org/10.1007/978-981-13-9235-1_5

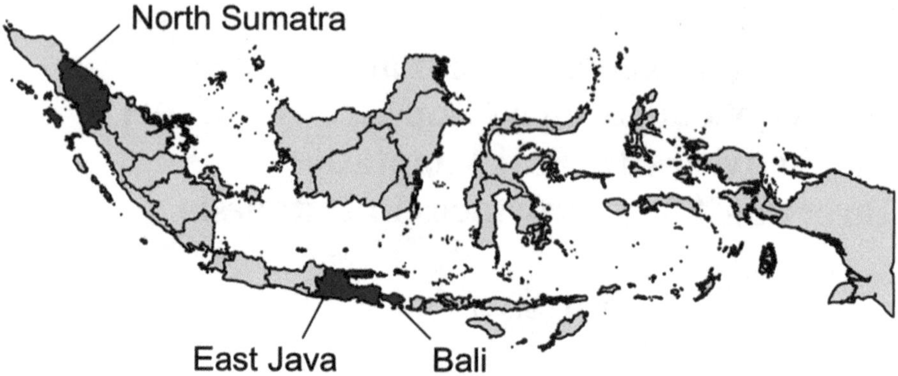

Fig. 5.1 Target provinces of Indonesia in the present study: North Sumatra, East Java, and Bali

increase in atmospheric CO_2 concentration (Pachauri et al. 2014), which possibly affects detrimental impacts on agricultural production (e.g., Rosenzweig and Parry 1994; Parry et al. 2004; Porter et al. 2014).

Rice is one of the major crops in the world, especially in Asia, wherein critical food shortage occurs (von Grebmer et al. 2017). At the lower latitudes in Asia, where the air temperature is already relatively high, a further increase in air temperature can reduce the rice yield (Parry et al. 2007; Rosenzweig et al. 2014). On the other hand, precipitation change is also likely to affect rice yield, because rainfed paddy fields are widely distributed in developing countries in Asia. Therefore, increase in the air temperature and precipitation change might particularly reduce the rice production in countries located at lower latitudes, despite the positive effects of increase in atmospheric CO_2 concentration on rice production.

Indonesia, located along the equator in Southeast Asia, is ranked as a country with the third largest rice production and sixth largest rice consumption per capita in the world (Maclean 2013). Recently, concerns have been expressed regarding the detrimental impact of climate change on the rice production in Indonesia. To mainstream the climate change adaptation issue into not only the national-level policy of Indonesia but also in the province-level policy, the National Climate Change Adaptation Action Plan (RAN-API) has been formulated in Indonesia. Furthermore, following the Paris agreement (UNFCCC 2015), the Ministry of the Environment (MOE) of Japan is working in collaboration with the Ministry of National Development Planning of the Republic of Indonesia (BAPPENAS) to enhance the capacity of policymakers as well as researchers through the cooperation for climate impact assessment and development of adaptation strategies in Indonesia. To achieve this, three provinces in Indonesia, North Sumatra, East Java, and Bali, were selected as some of the major provinces in Indonesia (Fig. 5.1).

A simulation model is a useful tool for forecasting the impacts of climate change on the rice yield. The impact assessment of climate change on agricultural productivity has been performed at the national level (e.g., Parry et al. 2004). However, more local impact assessment, for example, at the province and district levels, is required to cope with the detrimental impacts of climate change, because a reasonable strategy to counter the impact of climate change depends on the local heterogeneity of meteorological, geological, and socioeconomic characteristics, which are available in the local governments.

Recently, Masutomi et al. (2016a, b) developed a process-based crop growth model, MATCRO-Rice. In the present study, we introduced a case study on the impact assessment of climate change on rice productivity in three provinces in Indonesia using MATCRO-Rice. Furthermore, we evaluated the recent activities in Indonesia to enable the local government along with the local stakeholders to mainstream climate change adaptation.

5.2 Impact Assessment of Climate Change on Rice Productivity for Adaptation Planning

5.2.1 Overview of Crop Growth Model

A crop growth model, MATCRO-Rice, was used in the present study to simulate the climate change impacts on rice yield (Fig. 5.2). This model can simulate it from meteorological data by evaluating the influence of climate change on plant phenology and physiological characteristics such as photosynthesis, respiration, and allocation of photosynthetic products to each plant organ. Because the growth characteristics such as the allocation of photosynthetic products depend on the cultivar, the growth parameters were tuned for Ciherang, which is Indonesian major rice cultivar (USDA 2010), using experimental data conducted under the International Rice Research Institute (IRRI)-Japan Collaborative Research Project (IJCRP).

In the present study, we made some simplifications and modifications to MATCRO-Rice. To simplify the calculations due to enormous calculations, the leaf and soil temperatures were assumed to be the same as the air temperature. The CO_2

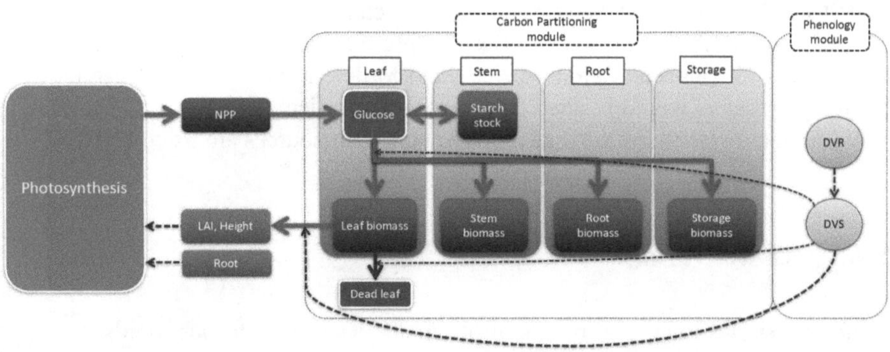

Fig. 5.2 Overview of a crop growth model, MATCRO-Rice, used in the present study (Masutomi et al. 2016a, b)

fertilization effect on the yield was considered by multiplying the simulated yield by a fertilizer coefficient (Tubiello et al. 2007). Maximum ribulose 1,5-bisphosphate carboxylase/oxygenase capacity at the canopy top ($V_{max}(0)$) for calculating photosynthetic rate was newly determined by specific leaf nitrogen (S_{LN}) as follows:

$$V_{max}(0) = \mathrm{MIN}\left(87.04 \times \left(S_{LN} - 0.487\right), 138.77\right).$$

S_{LN} was determined by the amount of nitrogen supplied to the soil (N_{FERT}) as follows:

When D_{VS}<X1:

$$S_{LN} = \mathrm{Y1}$$

When X1 < D_{VS} < X2:

$$S_{LN} = \left(\mathrm{Y1} - \mathrm{Y2}\right) \times D_{VS} / \left(\mathrm{X1} - \mathrm{X2}\right) + \left(\mathrm{Y2} \times \mathrm{X1} - \mathrm{Y1} \times \mathrm{X2}\right) / \left(\mathrm{X1} - \mathrm{X2}\right)$$

When X2 < D_{VS} < X3:

$$S_{LN} = \left(\mathrm{Y2} - \mathrm{Y3}\right) \times D_{VS} / \left(\mathrm{X2} - \mathrm{X3}\right) + \left(\mathrm{Y3} \times \mathrm{X2} - \mathrm{Y2} \times \mathrm{X3}\right) / \left(\mathrm{X2} - \mathrm{X3}\right)$$

where D_{VS} is developmental stage from planting date (D_{VS} = 0.0) to harvest date (D_{VS} = 1.0). X1, X2, X3, Y1, Y2, and Y3, crop-specific parameters, are 0.1, 0.2, 1.2, 0.774, 3.2 – (3.2–1.3) × $e^{-0.009 \times \mathrm{NFERT}}$, and 0.5, respectively.

5.2.2 *Impact Assessment Procedure*

To validate the yield predictability of the MATCRO-Rice tuned for Ciherang, we first compared the simulated yield to observe yield in each province reported by BPS (2016) and then simulated the impact of climate change on yield. We then performed a sensitivity analysis of the single effect of future air temperature, precipitation, or atmospheric CO_2 concentration on rice yield to determine which of these climate change parameters would be the key factor affecting rice yield change. A summary of the simulation settings, the data required, and their sources are listed in Table 5.1.

5.2.3 *Model Validation*

Figure 5.3 shows long-term trends of observed and simulated rice yields in North Sumatra, East Java, and Bali. In North Sumatra, the increasing trend in the observed yields were not predicted by MATCRO-Rice (r = –0.583). However, the absolute

Table 5.1 Simulation setting and model input

Parameter	Value or source
Plant	*Oryza sativa* cv. Ciherang
Base year (current)	1998–2002
Targeted year (future)	2018–2042
Calculation time step	3 hours
Spatial resolution	0.5° × 0.5° grid
Latitude	−12° to 7.5°
Longitude	94° to 142°
Current climate	WATCH-Forcing-Data-ERA-Interim (WFDEI) (Weedon et al. 2014)
Future climate	Four General Circulation Models (GCMs), BNU-ESM, CNRM-CM5, IPSL-CM5A-LR, and IPSL-CM5A-MR, from CMIP5 (Taylor et al. 2012) based on four Representative Concentration Pathways (RCP) scenarios (van Vuuren et al. 2011), with bias correction by Gomez-Garcia et al. (2017)
Atmospheric CO_2 concentration	Meinshausen et al. (2011)
Soil texture	Harmonized World Soil Database (FAO et al. 2012)
Cropping number and planting date	Laborte et al. (2017)
Harvested area	Sacks et al. (2010)
Nitrogen fertilization	Lu and Tian (2017)
Irrigation area	Global Map of Irrigation Areas (FAO 2016)
Photosynthetic response to drought	Bouman et al. (2001)
Yield response to CO_2	Tubiello et al. (2007)

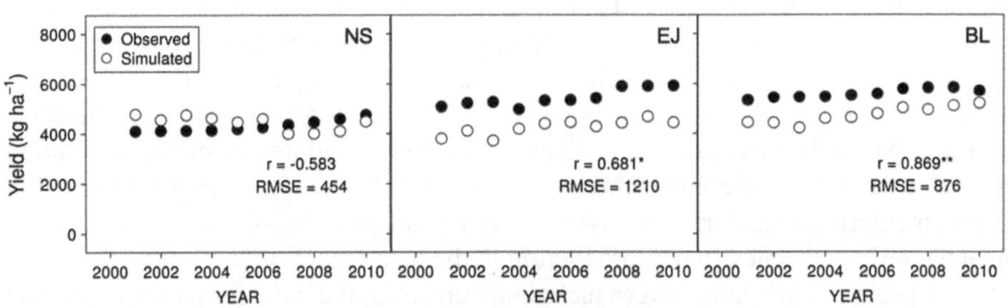

Fig. 5.3 Long-term trends in observed and simulated rice yield in North Sumatra (NS), East Java (EJ), and Bali (BL). RMSE: Root mean square error. ∗: $p < 0.05$, ∗∗: $p < 0.01$

yields simulated by MATCRO-Rice were almost the same as observed yields (RMSE = 454 kg ha^{-1}). In East Java and Bali, in contrast, the long-term trend in rice yield was similar in both the observations and the simulation ($r = 0.681$ and 0.869, respectively), but the absolute yields were different (RMSE = 1210 and 876 kg ha^{-1}, respectively). On the other hand, further analysis revealed that there is no significant relationship between the year-to-year changes of simulated yields and observed yields from the long-term yield trend (data not shown), which possibly indicates yield variability only due to climatic factor. As MATCRO-Rice could predicted interannual variability of rice yields in Japan (Masutomi et al. 2016b), this might be because we could not consider the annual change of modeling input data, for example, area ratio of irrigated paddy field to rainfed one and planting/harvest dates, which possibly changes annually depending on the climate and the timing of rainfall by monsoon onset (Naylor et al. 2007), respectively. Furthermore, farmers may have experienced effects of pests, diseases, and weed infestation on rice in a certain year. Actually, Li et al. (2015) reported that while the ensemble mean of simulated interannual variability in rice yield over the 13 crop models showed good agreement with the de-trended observed yields, each crop model could not reproduce the interannual variability. For these reasons, although we could not clarify whether MATCRO-Rice can truly predict climatic impacts on productivity of Ciherang to date, we tried to assess its impacts in the next chapter.

5.2.4 Impact Assessment of Climate Change

Several previous studies have simulated climate change impacts on rice yield (e.g., Parry et al. 2004). However, they targeted rice productivity at the national level, which is not so effective for making and implementing adaptation measures as the adaptation is needed at a more local, for example, provincial, level. Therefore, we focused on three provinces and districts in each province in our simulation of climate change impacts on rice yield. Figure 5.4 shows the changes in the rice yield due to climate change from 1998–2002 to 2019–2023, 2024–2028, 2029–2033, 2034–2038, or 2039–2042, according to the planning periods in the Regional Medium-Term Development Plan (*Rencana Pembangunan Jangka Menengah Desa* (RPJMD)). The simulation showed that the degrees of negative impact on rice yields will increase year by year in the three provinces. Therefore, creating and implementing adaptation plans for the impact of climate change on rice production in Indonesia are needed in the near future, particularly because the Indonesian Ministry of Agriculture aims to achieve self-sufficiency in rice production in the near future (IMoA 2016).

As there are limits in resources including human capital, land for paddy fields, and finances, prioritization is an important strategy in developing adaptation measures. In the present study, the negative impacts were highly dependent on the districts in each province, indicating the importance of regional adaptation priorities (Fig. 5.4). For local governments, regional heterogeneity in each province can help inform adaption planning. For example, the northern and eastern areas in North Sumatra place a high priority on countermeasures to climate change due to the high negative impacts on its rice yield.

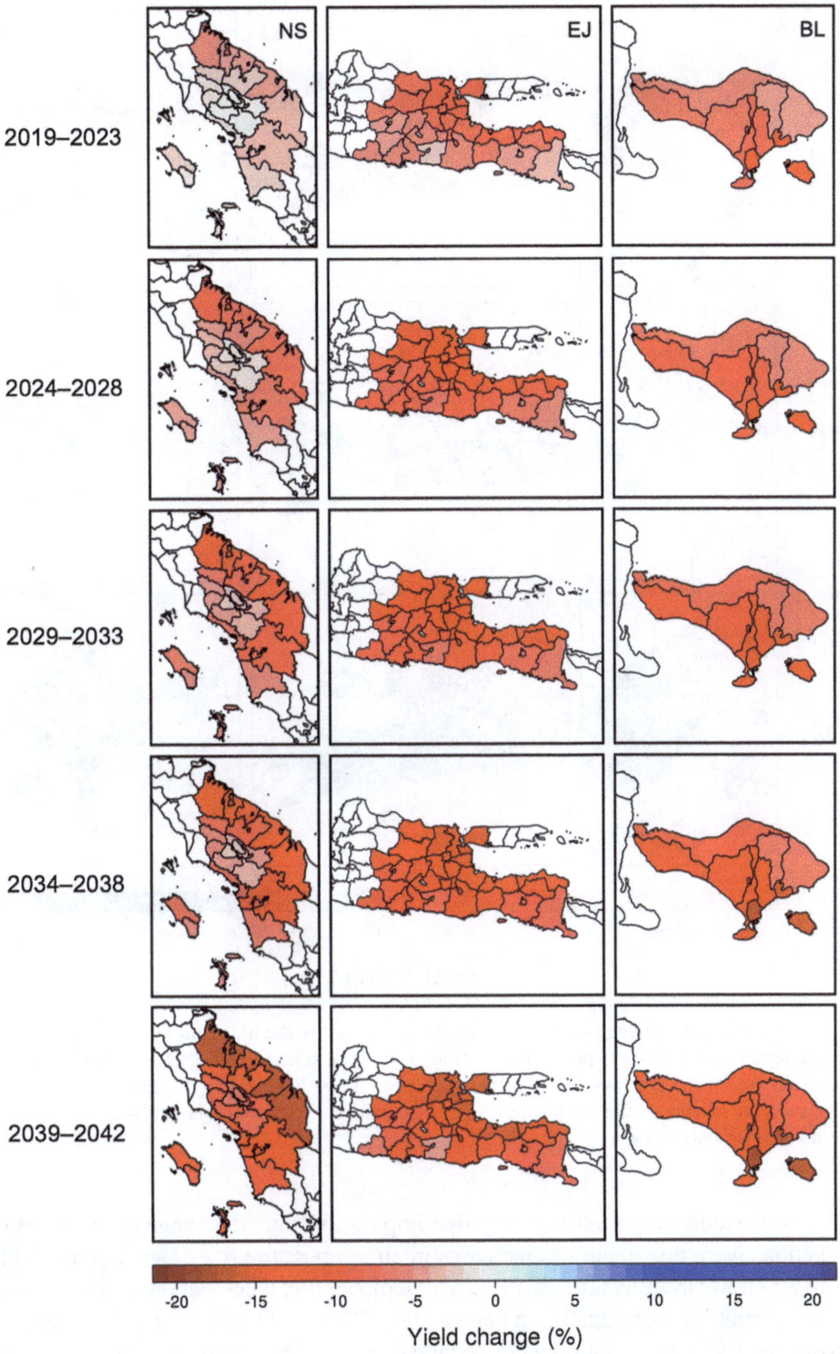

Fig. 5.4 Percent yield changes from 1998 to 2002 induced by climate change in North Sumatra (NS), East Java (EJ), and Bali (BL) averaged across the change in yield based on 14 future climate scenarios (a combination of four GCMs and four RCPs; there were no data for two scenarios)

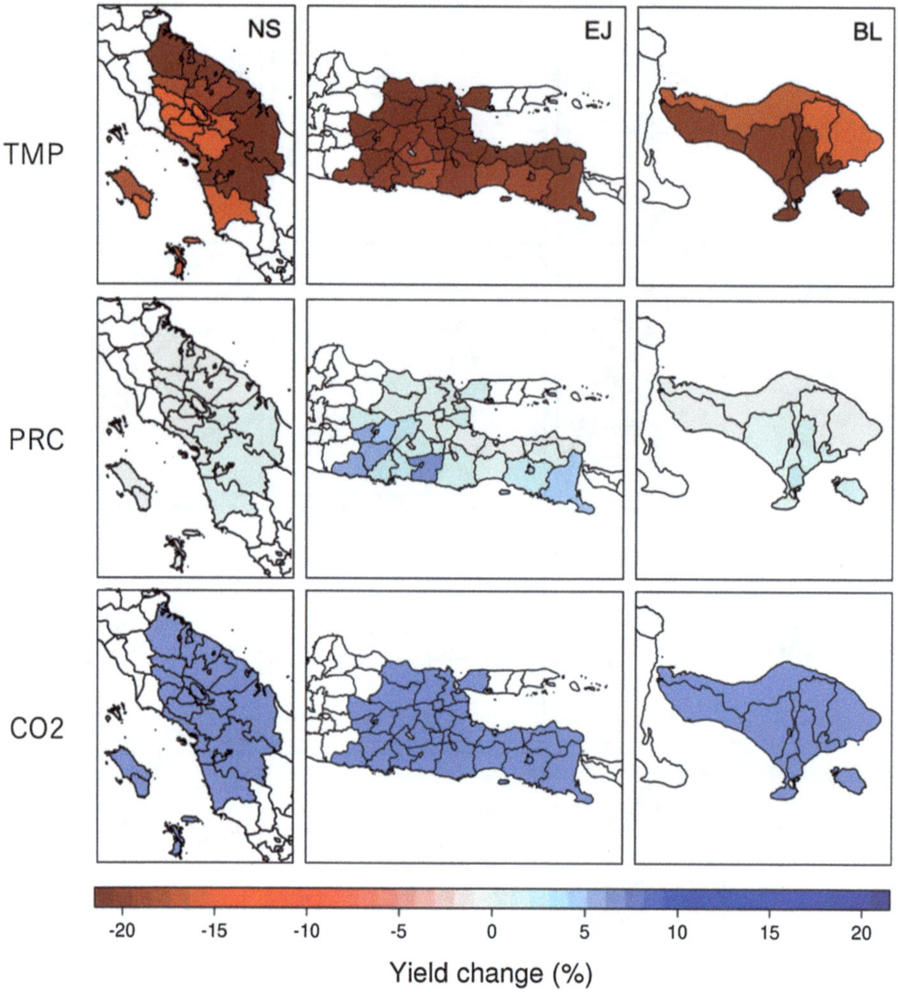

Fig. 5.5 Percent yield changes from 1998 to 2002 induced by the individual climate change factors, air temperature (TMP), precipitation (PRC), or atmospheric CO_2 concentration (CO_2) in North Sumatra (NS), East Java (EJ), and Bali (BL) for 2039–2042, averaged across the change in yield based on 14 future climate scenarios (a combination of four GCMs and four RCPs; there were no data for two scenarios)

Previous studies reported that negative impacts of climate change on agricultural productions were due mainly to increases in air temperature (e.g., Babel et al. 2011). Figure 5.5 shows the effects of changes in air temperature, precipitation, or atmospheric CO_2 concentration, individually, on rice yield for 2039–2042. It indicates that change in atmospheric CO_2 concentration increased the rice yield. In contrast, increase in air temperature lead to reductions in yield, which is consistent with the previous results (Babel et al. 2011). Therefore, breeding a new cultivar or screening a cultivar with high tolerance to high air temperature would be needed to adapt rice production to climate change.

5.2.5 Uncertainties in the Impact Assessment

Uncertainties in future climate scenarios could result in variations in the impacts of climate change on rice yield. Our simulations of the impacts of climate change on rice yield used each climate scenario from a total of 14 climate scenarios (a combination of four GCMs and four RCPs; there were no data for two scenarios), and those averaged impacts are shown in Figs. 5.4 and 5.5. On the other hand, Fig. 5.6 shows the variations in the impacts of climate change on rice yield obtained from 14 climate scenarios for 2039–2042. In each district of each province, the simulated yield change varies widely, approximately ±10%. However, the climate change-induced yield changes were negative in most districts for 2039–2042, even if the variations caused by climate scenarios were considered. Thus, the reductions in rice yield simulated for 2039–2042 are plausible in these districts.

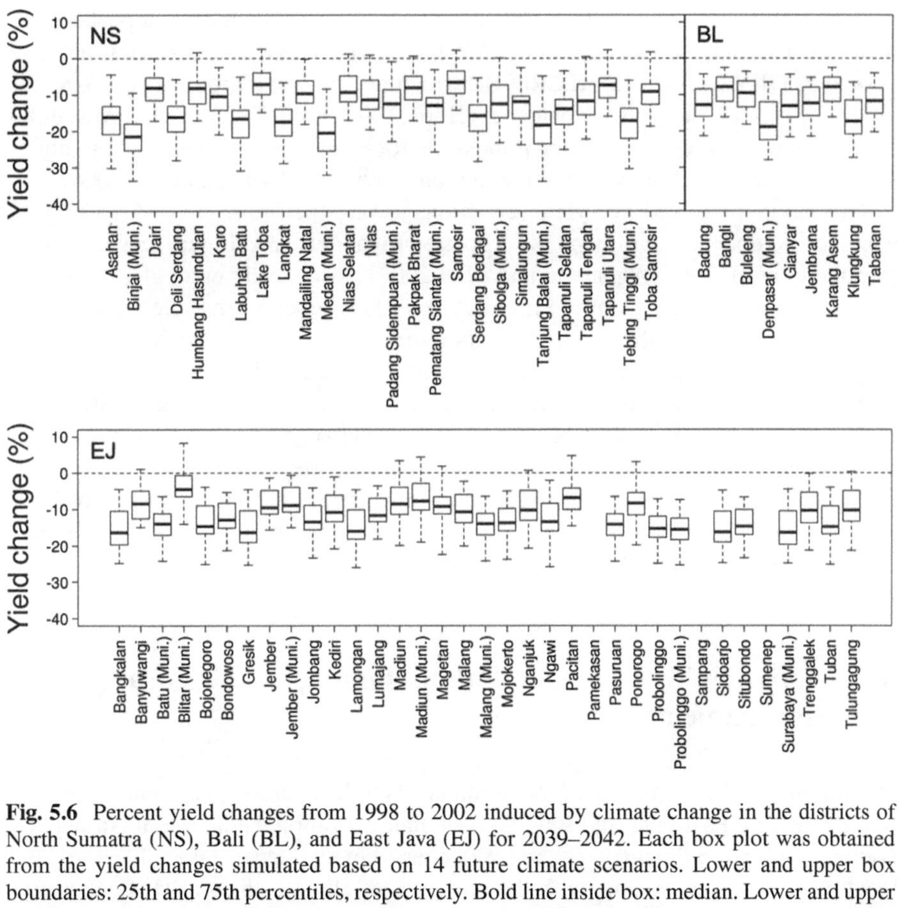

Fig. 5.6 Percent yield changes from 1998 to 2002 induced by climate change in the districts of North Sumatra (NS), Bali (BL), and East Java (EJ) for 2039–2042. Each box plot was obtained from the yield changes simulated based on 14 future climate scenarios. Lower and upper box boundaries: 25th and 75th percentiles, respectively. Bold line inside box: median. Lower and upper extreme of whisker: minimum and maximum values, respectively

Although our simulations showed yield reductions due to climate change in North Sumatra, East Java, and Bali, Matthews et al. (1997) obtained contrary results over all of Indonesia; they found that climate change had positive effects on rice yield. This difference in the impact assessments might be due to differences in the future climate scenarios used, which have a large impact on yield as can be seen from the Fig. 5.6. On the other hand, Matthews et al. (1997) reported that such differences of the climate impacts depend not only on future climate scenarios but also on crop growth models. They simulated the effects of climate change on rice production using two crop growth models, ORYZA1 and SIMRIW. They showed that the change in climate change-induced rice yield in Japan was −14.2% with ORYZA1 but +9.7% with SIMRIW when the same climate scenario, Geophysical Fluid Dynamics Laboratory (GFDL), was used. As a result of intercomparison of crop growth models (e.g., Li et al. 2015), it was clarified that difference in the simulated yields among models often depends on whether the model considers extreme temperature impacts on crop phenology and harvest index. In this perspective, MATCRO-rice, ORYZA1, and SIMRIW include its impacts on harvest index or spikelet fertility into the models. On the other hand, the crop model-dependent climate impacts may be caused by whether the models consider respiration process. Although ORYZA1 and MATCRO-Rice include the respiration process into the models, SIMRIW does not consider it (Horie 1987; Kropff et al. 1994; Masutomi et al. 2016a). As a result, it will be impossible for SIMRIW to reproduce high nighttime air temperature-induced yield reduction as reported by Peng et al. (2004).

Our model predicted rice yield reductions toward the future (Figs. 5.4 and 5.6), which could depend not only on the crop growth models and climate scenarios used (Matthews et al. 1997; Aggarwal and Mall 2002; Rosenzweig et al. 2014) but also on rice cultivars. For example, Prasad et al. (2006) reported that the yield of rice cultivar N-22 was not affected by high air temperatures in 2002, while the yields of 13 other cultivars decreased by degrees varying greatly from 48.2% to 94.9%. Nevertheless, several models have not targeted county-specific rice cultivars in global impact assessments of the effect of climate change on crops. The strength of our analysis was in our use of a model tuned to the major local rice cultivar "Ciherang" using field experimental data. Considering the characteristics of local crop cultivars could be critical in assessing the regional impacts of climate change on crop yield and in developing adaptation plans.

5.3 Activities Toward Adaptation Planning in the Local Government

Our findings indicate that rice production needs to be adapted to climate change in North Sumatra, East Java, and Bali. To formulate and implement an adaptation plan, not only national policymakers and researchers, but also local stakeholders and the local government must cope with the problems of climate change. This is because

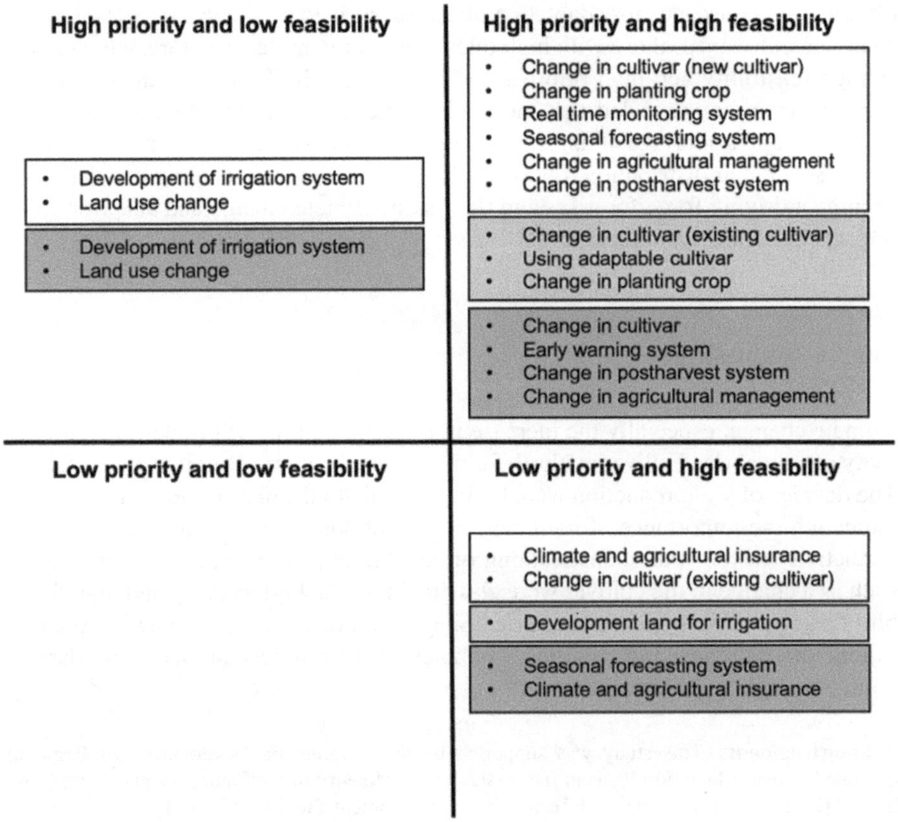

Fig. 5.7 Classification of adaptation options by the priority (importance and effectiveness for increasing crop production) and feasibility (easiness to implement the options from a viewpoint of financial and technical problems). First, our research group and local stakeholders individually suggested adaptation options (white and light gray, respectively). After a mutual discussion, a consensus regarding possible adaptation plans was arrived at (dark grey)

agricultural problems are particular to each province, and the data required to cope with the problem are only available locally in each province.

To support risk assessment in Indonesian national and local governments for climate change and the mainstreaming of the adaptation process, we conducted symposiums, capacity-building assistance, and workshops in North Sumatra, East Java, and Bali. Here, we introduce the results of a workshop conducted in February 2018. The objective was to forge agreements with local stakeholders regarding the adaptation plan. In the workshop, after the introducing our results on the impacts of climate change on rice yield, our research group and the local stakeholders individually classified adaptation strategies into four groups (Fig. 5.7): "low priority and low feasibility," "high priority and low feasibility," "low priority and high feasibility," or "high priority and high feasibility." Through the mutual discussion on this classification considering the information we have and local agricultural problems together, ultimately, a consensus for adaptation plans was arrived at. Some strategies, such as changing cultivars to those with high tolerance to high air temperature, were ranked as high feasibility and high priority, and these were similar among the provinces. Further discussion revealed that the main challenge to the adaptation is financial deficiency and a shortage of databases and/or information regarding cultivar characteristics and their distribution routes. We hope our activities including the workshop become a driving force for assessing the risk of climate change and adaption to its impacts in national and local levels.

5.4 Conclusions

Climate change, especially the increase in air temperature, will impair the productivity of a local rice cultivar in North Sumatra, East Java, and Bali in the near future. The degrees of yield reduction were highly specific to the districts in each province, indicating the importance of regional priority for adaptation to climate change. In the activity toward the mainstreaming of adaptation processes, adaptation options such as a change in the cultivar were classified into the high priority and high feasibility category. We hope the mainstreaming the climate change adaptation issue in Indonesian national- and local-level policies, and local stakeholders can adapt to climate change in rice production.

Acknowledgments This study was supported by the Program on Development of Regional Climate Change Adaptation Plans in Indonesia, by the Ministry of the Environment of Japan, and by the Environment Research and Technology Development Fund (S-12) of the Environmental Restoration and Conservation Agency. This study was performed in collaboration with the Indonesian Ministry of National Development Planning (BAPPENAS; Badan Perencanaan Pembangunan Nasional). We are indebted to Dr. Keiichi Hayashi for providing us with experimental data, from studies conducted in IJCRP, which were funded by the Ministry of Agriculture, Forestry, and Fisheries of Japan, regarding tuning the crop growth model. We acknowledge the World Climate Research Programme's Working Group on Coupled Modelling, which is responsible for CMIP. We also thank the climate modeling groups (listed in Table 5.1) for producing and making available their model output. For CMIP, the U.S. Department of Energy's Program for Climate Model Diagnosis and Intercomparison provided coordinating support and led the development of software infrastructure in partnership with the Global Organization for Earth System Science Portals.

References

Aggarwal PK, Mall RK (2002) Climate change and rice yields in diverse agro-environments of India. II. Effect of uncertainties in scenarios and crop models on impact assessment. Clim Chang 52:331–343

Babel MS, Agarwal A, Swain DK et al (2011) Evaluation of climate change impacts and adaptation measures for rice cultivation in Northeast Thailand. Clim Res 46:137–146

Bouman BAM, Kropff MJ, Tuong TP et al (2001) In: International Rice Research Institute, and Wageningen: Wageningen University and Research Centre (ed) ORYZA2000: modeling lowland Rice, Los Baños (Philippines)
BPS (Badan Pusat Statistik) (2016) Food crops. https://www.bps.go.id/subject/53/tanaman-pangan.html#subjekViewTab3. Accessed 30 November 2018
Dlugokencky E, Tans P (2018) NOAA/ESRL. http://www.esrl.noaa.gov/gmd/ccgg/trends/. Accessed 30 November 2018
FAO (Food and Agriculture Organization of the United Nations) (2016) Global Map of Irrigation Areas (GMIA) 2013, AQUASTAT. http://www.fao.org/nr/water/aquastat/irrigationmap/index10.stm. Accessed 30 November 2018
FAO (Food and Agriculture Organization of the United Nations), IIASA (International Institute for Applied Systems Analysis), ISRIC (International Soil Reference and Information Centre), ISSCAS (Institute of Soil Science, Chinese Academy of Sciences), JRC (Joint Research Centre) (2012) Harmonized World Soil Database (version 1.2)
Gomez-Garcia M, Ogawada D, Matsumura A et al (2017) Evaluation of the application of the ISI-MIP bias-correction method of future simulations of climate over Indonesia for the implementation of climate change adaptation plans. HESSS4 Tokyo, Japan
Horie T (1987) A model for evaluating climatic productivity and water balance of irrigated rice and its application to Southeast Asia. Southeast Asian Stud 25:62–74
IMoA (Indonesian Ministry of Agriculture) (2016) Rencana Strategis Kementerian Pertanian Tahun 2015–2019. http://sakip.pertanian.go.id/admin/file/Renstra%20Kementan%202015-2019%20(edisi%20revisi).pdf. Accessed 30 November 2018 (In Indonesian)
Joos F, Spahni R (2008) Rates of change in natural and anthropogenic radiative forcing over the past 20,000 years. Proc Natl Acad Sci USA 105:1425–1430
Kropff MJ, van Laar HH, Matthews RB (1994) ORYZAl: an ecophysiological model for irrigated rice production. In: SARP research proceedings, AB-DLO, Wageningen, The Netherlands
Laborte AG, Gutierrez MA, Balanza JG et al (2017) RiceAtlas, a spatial database of global rice calendars and production. Sci Data 4:170074
Li T, Hasegawa T, Yin X et al (2015) Uncertainties in predicting rice yield by current crop models under a wide range of climatic conditions. Glob Chang Biol 21:1328–1341
Lu C, Tian H (2017) Global nitrogen and phosphorus fertilizer use for agriculture production in the past half century: shifted hot spots and nutrient imbalance. Earth Syst Sci Data 9:181–192
Maclean JL (2013) Rice almanac: source book for the most important economic activity on earth. International Rice Research Institute
Masutomi Y, Ono K, Mano M et al (2016a) A land surface model combined with a crop growth model for paddy rice (MATCRO-Rice v. 1)-part 1: model description. Geosci Model Dev 9:4133–4154
Masutomi Y, Ono K, Takimoto T et al (2016b) A land surface model combined with a crop growth model for paddy rice (MATCRO-Rice v. 1)-part 2: model validation. Geosci Model Dev 9:4155–4167
Matthews RB, Kropff MJ, Horie T et al (1997) Simulating the impact of climate change on rice production in Asia and evaluating options for adaptation. Agric Syst 54:399–425
Meinshausen M, Smith SJ, Calvin K et al (2011) The RCP greenhouse gas concentrations and their extensions from 1765 to 2300. Clim Chang 109:213–241
Naylor RL, Battisti DS, Vimont DJ et al (2007) Assessing risks of climate variability and climate change for Indonesian rice agriculture. Proc Natl Acad Sci USA 104:7752–7757
Pachauri RK, Allen MR, Barros VR et al (2014) Climate change 2014: synthesis report. In: Pachauri R, Meyer L (eds) Contribution of working groups I, II and III to the fifth assessment report of the Intergovernmental panel on climate change, IPCC, Geneva, Switzerland
Parry ML, Rosenzweig C, Iglesias A et al (2004) Effects of climate change on global food production under SRES emissions and socio-economic scenarios. Glob Environ Chang 14:53–67

Parry ML, Canziani OF, Palutikof JP et al (2007) Climate change 2007: impacts adaptation and vulnerability. Contribution of working group II to the fourth assessment report of the intergovernmental panel on climate change. Cambridge University Press, Cambridge
Peng S, Huang J, Sheehy JE et al (2004) Rice yields decline with higher night temperature from global warming. Proc Natl Acad Sci USA 101:9971–9975
Porter JR, Xie L, Challinor AJ et al. (2014) Food security and food production systems. In: Field CB, Barros VR, Dokken DJ, Mach KJ, Mastrandrea MD, Bilir TE, Chatterjee M, Ebi KL, Estrada YO, Genova RC, Girma B, Kissel ES, Levy AN, MacCracken S, Mastrandrea PR, White LL (eds) Climate change 2014: impacts, adaptation, and vulnerability. Part a: global and sectoral aspects. Contribution of working group II to the fifth assessment report of the intergovernmental panel on climate change. Cambridge University Press, Cambridge/New York, pp 485–533
Prasad PVV, Boote KJ, Allen LH Jr et al (2006) Species, ecotype and cultivar differences in spikelet fertility and harvest index of rice in response to high temperature stress. Field Crops Res 95:398–411
Rosenzweig C, Parry ML (1994) Potential impact of climate change on world food supply. Nature 367:133–138
Rosenzweig C, Elliott J, Deryng D et al (2014) Assessing agricultural risks of climate change in the 21st century in a global gridded crop model intercomparison. Proc Natl Acad Sci USA 111:3268–3273
Sacks WJ, Deryng D, Foley JA et al (2010) Crop planting dates: an analysis of global patterns. Glob Ecol Biogeogr 19:607–620. https://nelson.wisc.edu/sage/data-and-models/crop-calendar-dataset/index.php. Accessed 30 November 2018
Taylor KE, Stouffer RJ, Meehl GA (2012) An overview of CMIP5 and the experiment design. Bull Am Meteor Soc 93:485–498
Tubiello FN, Amthor JS, Boote KJ et al (2007) Crop response to elevated CO2 and world food supply: a comment on "Food for Thought..." by Long et al., Science 312:1918–1921, 2006. Eur J Agron 26:215–223
UNFCCC (United Nations Framework Convention on Climate Change) (2015) The Paris Agreement. http://unfccc.int/paris_agreement/items/9485.php. Accessed 30 November 2018
USDA (United States Department of Agriculture) (2010) Gain report. https://gain.fas.usda.gov/Recent%20GAIN%20Publications/Rice%20and%20Corn%20Update_Jakarta_Indonesia_4-23-2010.pdf. Accessed 30 November 2018
van Vuuren DP, Edmonds J, Kainuma M et al (2011) The representative concentration pathways: an overview. Clim Chang 109:5–31
von Grebmer K, Bernstein J, Hossain N et al. (2017) 2017 Global hunger index: the inequalities of hunger. http://www.ifpri.org/cdmref/p15738coll2/id/131422/filename/131628.pdf. Accessed 30 November 2018
Weedon GP, Balsamo G, Bellouin N et al (2014) The WFDEI meteorological forcing data set: WATCH forcing data methodology applied to ERA-interim reanalysis data. Water Resour Res 50:7505–7514

Chapter 6
Physiological Disorders and Their Management in Greenhouse Tomato Cultivation at High Temperatures

Katsumi Suzuki

Abstract In a greenhouse, an artificial environment is created inside a covering and crops are produced therein so as to cope with changes in the weather. However, cooling is more difficult to achieve than heating, and it is thus necessary to develop technology that addresses high temperatures. Tomato is a major vegetable grown in greenhouses. In tomato cultivation under high-temperature conditions, pollen fertility decreases, fertilization fails, and poor fruit sets result in decreased fruit yields. The incidence of physiological disorders (e.g., fruit cracking, blossom-end rot, leaf marginal necrosis, and yellow shoulder fruit) also increases. The mechanism underlying the occurrence of each physiological disorder and management techniques for suppressing these disorders are introduced in this chapter. Heat countermeasure technologies used in greenhouses in Japan, such as night cooling using a heat pump, fog cooling, and the use of appropriate covering materials, are also explained.

6.1 Introduction

Changes in the weather have a great influence on the production of open-field crops. The risk from global warming is also substantial for protected horticulture. The interior of greenhouses in Japan during summer has recently become dangerously hot. Ventilation is the primary means for decreasing the air temperature in greenhouses, but as temperatures rise, it will be increasingly difficult to manage air temperature by ventilation alone. Under high-temperature conditions, there is increased risk of various physiological disorders that do not pose a problem under optimal temperatures. Abnormal weather will continue due to meteorological fluctuations in the future, and it is necessary to develop new technologies for greenhouses to address excessive high temperatures and meteorological fluctuations.

K. Suzuki (✉)
Faculty of Agriculture, Shizuoka University, Shizuoka, Japan
e-mail: suzuki.katsumi@shizuoka.ac.jp

T. Iizumi et al. (eds.), *Adaptation to Climate Change in Agriculture*,
https://doi.org/10.1007/978-981-13-9235-1_6

In greenhouse horticulture, an artificial environment is created inside plastic films or glass coverings, and crops are produced therein, providing some measure of protection against weather fluctuations. There are several cost-effective ways to heat the inside of a greenhouse when outside air temperature is low. In contrast, lowering air temperatures inside a greenhouse is a major challenge. Various problems thus arise in greenhouses during high-temperature periods.

In Japan, many vegetables are produced in greenhouses, but there are problems with each crop during summer high temperatures. Physiological disorders that occur in high temperatures include tip burn in lettuce, leaf senescence in leaf onion, stem elongation in Chinese cabbage, decrease of fruit set in some fruit vegetables (e.g., flower drop in snap bean), fruit color disorder (e.g., eggplant), and fruit shape disorder (e.g., eggplant and cucumber). Although there are various horticultural crops that are sensitive to high temperatures, this chapter focuses on tomatoes. Tomato is one of the most important crops produced in greenhouses, and many studies on high-temperature injuries and physiological disorders have been published. In addition, various countermeasure technologies against such negative influences are also being developed.

Tomatoes, of which the agricultural output value is highest among vegetables in Japan, are supplied year-round. Tomato production is affected by high temperatures and the yield, particularly of good-quality fruits, is decreased. In Japan, to avoid excessive temperatures, tomato fruits are often produced in cool areas, that is, high-altitude mountainous areas and high-latitude regions such as Hokkaido and Tohoku, even though it remains hot during the summer season. The main reasons for the decreased yield of good-quality tomato fruit due to higher temperatures are poor fruit sets, reduced plant vigor, and development of physiological disorders. Recently, the number of large-scale greenhouses with multiple employees has been increasing in Japan. In a large number of such greenhouses, tomatoes are produced annually, and management techniques to combat high temperatures are an important issue.

In tomato cultivation under high-temperature conditions, the incidence of physiological disorders in fruits, such as fruit cracking, blossom-end rot, and yellow shoulder fruits, has increased. These physiological disorders are not necessarily attributable to high temperatures, but they readily occur in tomato cultivation during the summer, causing problems in production. As the best countermeasure is to lower the air temperature inside the greenhouse, we will also discuss the cooling technologies used in greenhouses in Japan. In summer tomato cultivation, biological stressors, such as yellow leaf roll disease, whitefly, thrips, leaf mold disease, and ticks, are serious problems. This chapter will not deal with biotic stress, but will instead focus on physiological disorders caused by abiotic stress. In many cases, the disorders can be solved by breeding, for example, development of disease-resistant cultivars, hard-to-crack cultivars, strong vigor cultivars, parthenocarpic cultivars, etc. As the breeding of heat-tolerant crops is an important task, there is a significant amount of genetic research being conducted (Ainsworth and Ort 2010; Driedonks et al. 2016). However, this chapter will not address breeding issues or technology.

6.2 Plant Damage Due to High Temperature

6.2.1 Decreased Pollen Fertility

In tomato plants, the aspect that is the most sensitive to damage from high temperatures is the development of pollen (Paupière et al. 2017). The pollen development process in many other plant species is also damaged by high temperatures (Rieu et al. 2017). In summer, even in high-altitude cool areas, air temperature inside a greenhouse is often over 30 °C. In such cases, the pollen becomes sterile, fertilization fails, and the fruit does not set. Therefore, in Japan, it is common for fruit setting to use 4-CPA (artificial auxin:4-chlorophenoxyacetic acid) in order to deal with high temperature and cause parthenocarpic fruit.

If a farmer judges that the pollen fertility has decreased at high temperatures, plant growth regulators are used. A mixed solution of gibberellin and auxin has been shown to improve the fruit set rate, even under high temperatures in summer (Bünger-Kibler and Bangerth 1983; Sasaki et al. 2005).

Farmers commonly use bumblebees and Japanese black bumblebees for fruit setting. Their bee nests are also damaged under high temperatures, and thus a cooler that is dedicated to the nest box is used in the summer.

6.2.2 Decreased Plant Vigor

Plant vigor is a judgment index based on factors such as growth rate, sink–source balance, stem diameter, or the sum of these factors. "Plant vigor" is not an academic term, but it is often used in tomato production. In general, plant vigor is advanced at high temperatures and receded at low temperatures. Raising temperature leads to an increase in respiration rate and a decrease in photosynthesis. However, as the growth rate increases, the plant vigor declines. Fruit size decreases when the plant vigor is low, and some trusses have no fruit when the plant vigor is very low. For this reason, in simple rain-proof cultivation, farmers control the fruit set number in each truss for maintaining the plant vigor due to fruit limitation.

It is also effective to graft a target cultivar onto a strong rootstock. Plants grafted on a strong rootstock are more vigorous than self-rooted plants. The plant vigor can be maintained, potentially resulting in increased yield.

6.3 Environmental Control Technology for Cooling in Summer in Greenhouses

By using technology to overcome high temperatures, it is possible to solve the problems that develop at high temperatures in a greenhouse. In the greenhouses used for tourism in Japan, for example, air temperature inside the greenhouses is controlled by air conditioning units with sufficient cooling capacity, so that the greenhouses are suitable for crop cultivation all year round. In order to cool the inside of the greenhouse with such air conditioners, the initial and running costs are very large, which is not realistic for those producing vegetables as food. In many cases in crop production, therefore, it is not cost-effective to use air conditioning, particularly during the day.

6.3.1 Use of a Heat Pump

In Japan, the use of heat pumps for heating is increasing as an energy-saving technique to address the increase in oil prices. As it is also possible to provide cooling with a heat pump, the use of heat pumps during the summer to decrease the air temperature inside greenhouses is desirable, in part because it contributes to the effective use of this equipment. Efficient cooling methods using heat pumps equipped for heating have been tested, because it can be very difficult to reduce the air temperatures inside greenhouses during the day in summer.

Night cooling is a realistic means to use a heat pump as a cooler. Ohishi and Moriya (2011) indicated that the yield of tomatoes increased two- to threefold when they used a heat pump during the night in a greenhouse with multilayer covering material. The low night air temperatures contributed to an increase in the fruit set rate and suppression of fruit cracking. Iwasaki et al. (2011) showed that the fruit yield was significantly higher by night cooling with a heat pump. We investigated whether the tomato yield would increase with nighttime cooling by a heat pump and by early-morning CO_2 enrichment. The CO_2 concentration in the tested greenhouse increased mainly due to respiration from tomato plants (Fig. 6.1), and thus the CO_2 enrichment system (set to be activated at $\leq$600 μmol mol^{-1}) did not operate early in the morning. The fruit yield was increased by nighttime cooling in our examination (Fig. 6.1). This result may also be attributable to both the effect of nighttime cooling and the increase in CO_2 concentration in the greenhouse in the early morning.

Energy-saving use of a heat pump was investigated by Kawasaki and Ahn (2015): they carried out local cooling only around the shoot apex and flowers, which are sensitive to high temperatures, with a heat pump. Improved pollen fertility was observed when local cooling only near the shoot apex and flowers was performed at night. Furthermore, increased yield due to an increase in individual-fruit weight and the promotion of fruit enlargement was reported. As mentioned earlier, lowering the nighttime temperature in a greenhouse by operating a heat pump and lowering the

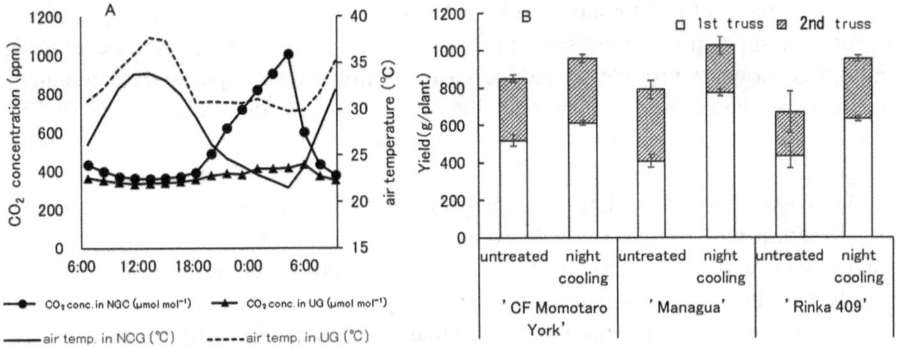

Fig. 6.1 (**a**) CO_2 concentration (ppm) and air temperature (°C) in night cooling greenhouses (NCG) and untreated greenhouses (UG). (**b**) Tomato fruits yield in three cultivars in night cooling greenhouses and untreated greenhouses

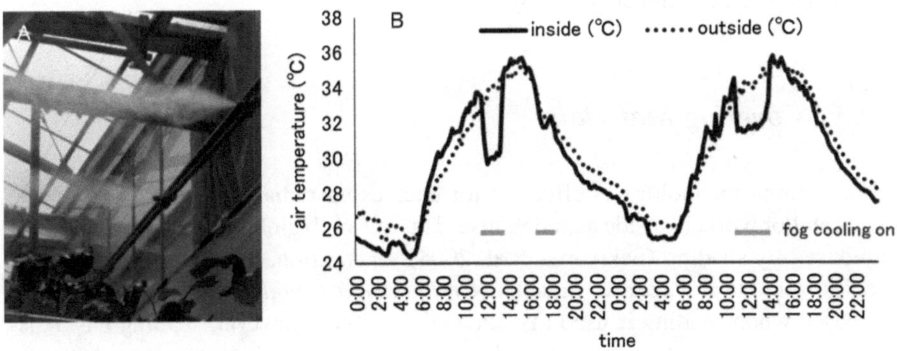

Fig. 6.2 (**a**) Fog-cooling system. (**b**) Changes in air temperatures inside and outside a greenhouse equipped with fog-cooling system (July 13-14, 2004)

temperature of sensitive organs in the plants are effective for avoiding injuries and disorders caused by high temperatures. Although the effects of using heat pumps in this manner and the calculation of the introduction and running costs have been examined, there have been few tests on a commercial scale. It is necessary to develop effective methods for local cooling only around the shoot apex and the flowers in actual cultivation.

6.3.2 Fog Cooling

Air temperature inside a greenhouse can also be lowered by evaporative cooling using the fog-cooling method (Fig. 6.2). However, the humidity is also raised by this method, so that a window(s) and/or a door must be opened for ventilation, at which time the air temperature reduction effect is no longer sustained. The effects of fog cooling on the yield of tomatoes at high temperatures in summer were also tested recently. A super-micro-mist spraying with a finer grain of water was developed, and it was shown to provide a yield increase; the fine mist is vaporized immediately after leaving the nozzle because the water grain was very fine (Ikeuchi et al. 2012). In an Aichi Prefecture experiment, air temperature in such super-micro-mist spraying greenhouses was >5 °C lower than that in conventional greenhouses (Hiei et al. 2015). Then, the super-micro-fog spraying method with shading was applied to tomato retarding culture, in which plants were planted on August 8. They reported that early growth was promoted and that yield within the year increased by approximately 14% compared to shading alone.

Fog cooling is now used as a technique that controls the humidity in an integrated environment control system, rather than simply lowering the air temperature. The opening of the stomata of tomato leaves promotes gas exchange and contributes to increasing net photosynthetic rate. As the extent of stomatal opening tends to be increased by high humidity, fog cooling has been used for the humidity control in greenhouses (Kinomoto et al. 2013).

6.3.3 Covering Materials

Light-shading technology is effective for decreasing radiative heat transfer to the plants and/or workers inside a greenhouse. The wet-bulb globe temperature (WBGT) is reduced by shading (Sakiyama et al. 2004). In addition, the yields of some crop species, such as leaf lettuce, mitsuba (*Cryptotaenia canadensis*), and spinach, are increased when shading is used (Yoshida et al. 2008). However, shading has a disadvantage for plants that require a relatively high level of light for photosynthesis, such as tomato and cucumber. Infrared-ray blocking films (IRB) that transmit lower amounts of radiation in the near-infrared wavelength with high thermal energy have been developed (Mitsubishi Chemical Agri Dream Co.). It has been shown that by using this film, the temperature of the plant surface is lowered. On the other hand, because the ratio of red to far-red light (the R/FR ratio) can also be increased, cell elongation is controlled or plant height is shortened. The temperatures of fruits remained low by IRB film-covering in watermelon, and the mean cell lengths of IRB were approximately 7–20% shorter than no-covering fruit (Fuluoka et al. 2009). In low eave greenhouses, the height of staking of tomato is low. If the stem length between trusses decreases, the number of trusses on the plant can increase. Sometimes, however, plant height shortening may have a negative effect from the aspect of photosynthesis with shielding of each leaf. This film is used in accordance with the purpose of each farmer for cultivation.

In recent years in Japan, universities and companies have been conducting joint research projects to develop highly coated materials. Among these materials, films that reflect infrared rays between about 800 and 1100 nm have been developed as a countermeasure to high temperatures. These films are expected to be marketed promptly.

6.4 Physiological Disorders

6.4.1 Fruit Cracking

The most significant problem in tomato production under high summer temperature conditions is fruit cracking (Dorais et al. 2004; Suzuki 2010). Fruit cracking is a physiological disorder in which the fruit surface cracks. There are three types of fruit cracking: radial fruit cracking (RFC) in which a deep crack appears radially from a calyx; cuticular cracking in which the surface cracks finely and becomes scratched; and skin cracking in which the pericarp surface cracks just before harvesting (Fig. 6.3). Among these, RFC is the most problematic.

The occurrence of RFC is also a major disorder in summer and autumn tomato production under rain shelter in cool upland areas. RFC occurs during fruit development, and thereafter, the crack spreads downward from the calyx as the fruit develops (Fig. 6.3). Fruits with serious RFC are no longer sellable and are discarded. Microscopic observations of the tissue after the occurrence of RFC show that damage occurs inside the pericarp cells rather than between cell walls (Fig. 6.3). It is thought that RFC is a result of physical destruction by an inflow of excess water and photosynthetic products into the fruit, so that the cell structure of the pericarp cannot be maintained.

In a study investigating the relationship between air temperature during fruit development and the occurrence of RFC, it was found that the RFC was more

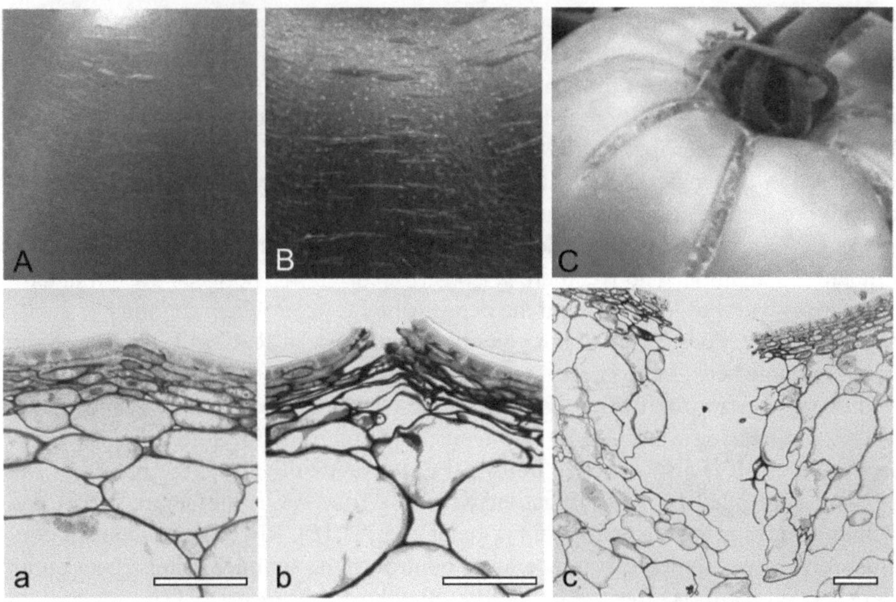

Fig. 6.3 Photographs and photomicrographs of cracking (bars in a, b, and c: 100 μm). (**a**, *a*) Thin cuticular cracking; (**b**, *b*) Thick cuticular cracking; (**c**, *c*) Radial fruit cracking

closely related to the average air temperature of the early fruit development than the average air temperature of the entire fruit developmental process (Sano et al., unpublished); during this early stage, cell division commonly occurs in the fruit and the cell number is increasing (Azzi et al. 2015). The rate of cell division decreases at high temperatures, and the cell number decreases when compared to low temperatures. When photosynthetic products and water are actively imported into the fruit, in which the cell number is small, during the growing period, the turgor pressure applied to each cell increases and the fruit becomes easy to crack. Molecular biological analyses have also shown that there is a positive correlation between the expression of genes involved in cell division and the occurrence of cracks (Ueno et al., unpublished).

Therefore, in order to prevent RFC, it is important that the initial cell division proceeds sufficiently and the number of fruit cells increases. RFC was reduced by shading fruits (Suzuki 2010). It is thought that the temperature of the fruit surface is lowered by sunlight shading, increasing the number of cells. Covering fruit with leaves to avoid direct sunshine irradiation on the fruit was reported to be effective at decreasing RFC in summer tomato production (Suzuki 2010).

Matsuo et al. (2012) suggested that cytokinins are involved in cell division during the development of tomato fruit. In order to increase the number of cells in early tomato fruits, the effect of an artificial cytokinin CPPU (1-(2-chloro-4-pyridyl)-3-phenylurea) was investigated. The marketable fruit yield was increased along with the reduction of RFC by spraying CPPU (Sano et al. 2018). The shape and quality of fruits sprayed with CPPU and those of untreated fruits were equivalent. The reduction of RFC by CPPU spraying was considered to be attributable to the increase in cell numbers near the fruit epidermis. CPPU spraying did not significantly affect the harvested fruit number, average fruit weight, or total fruit yield. In Japan, it is necessary to register plant growth regulators (PGR) as agricultural chemicals. CPPU was originally registered as a PGR for fruit trees, but not for tomatoes. Because it was clear that CPPU was able to decrease RFC, the use of CPPU was registered in December 2016 (Sumitomo Chemical Co.), and it can now be used in tomato production greenhouses.

As CPPU is expensive, it is desirable to spray it when the extent of RFC is significant. As mentioned above, there is a positive correlation between the initial temperature of fruit development and the degree of RFC occurrence, and thus, if the air temperature exceeds 25 °C in the greenhouse, the risk of RFC will increase (Sano et al., unpublished). In this case, spraying CPPU solution would be effective.

For suppression of RFC, other techniques, such as shading the whole greenhouse, suppressing water inflow into fruits, and controlling humidity by fog cooling, are used. It is also effective to use RFC-tolerant cultivars. If the humidity can be well controlled by humidification with fine fog (Aichi Prefecture 2012) and dehumidification with a heat pump (Ohishi et al. 2015), RFC can be reduced considerably. It was shown that temperature control in the summer using a heat pump as described above is effective for suppressing the occurrence of RFC.

In the absence of a system that controls humidity, it is effective to manage the sink–source balance. It is important to maintain the fruit number in the summer and autumn. Although the yield itself must be reduced in order to limit the amount of assimilation products that flow into the fruit, it is possible to reduce the occurrence of RFC by decreasing the source-to-sink ratio through shading or defoliation. Light shading reduces the photosynthesis rate, and the inflow of photosynthetic products into the fruits is reduced. Approximately 60–85% of the water to the fruit with photosynthetic products is transmitted by the phloem (Guichard et al. 2005). By suppressing the amount of photosynthetic products and water flowing into the fruit, the occurrence of RFC is reduced. Physical treatment such as tender press of the peduncle is also used to achieve the same effect (Suzuki et al. 2012).

The use of a ultraviolet (UV)-cut film that absorbs UV light is effective for reducing cuticular cracks (Kimura et al. 2012). This type of cracking can be reduced by approx. 60–70% by combining such film with irrigation control (Uetani et al. 2014). The UV-cut film is often used for a simple rain shelter for summer and autumn tomato production in Japan. The UV-cut film was also suggested to be effective for pest control (Ohta and Takeda 2014). A UV-cut film may influence bee pollination, but if direct light is partially transmitted instead of a complete covering, bees can remain active despite the film.

The occurrence of cracking skin is related to the magnitude of the vapor pressure deficit (VPD). Skin cracking is likely to occur when there is a large difference in the VPD between a day and one day before that day, and unlikely to occur when the difference is small (Watanabe et al. 2006). The control of day and night humidity is thus effective for suppressing the development of cracking skin.

6.4.2 Blossom-End Rot

In blossom-end rot (BER) fruits, the surface at the end of fruits turns black at the early developmental stage. The phenomenon that turns the inside brown or black is called internal browning (IB) (Suzuki et al. 2019). Blossom-end browning is when the epidermis and subepidermal layers turn brown in middle or late stages. All of these phenomena are caused by cell death inside the fruit (Fig. 6.4). These physiological disorders are induced by calcium (Ca) deficiency. Differences in these three types of injuries in tomato fruits with Ca deficiency are based on the stage and part at which Ca deficiency occurs (Fig. 6.4). BER and IB may occur at the same time, which may make them difficult to distinguish.

At the early stage of BER, the surface of the fruit changes to a water-immersion state and a black dot appears; eventually the entire surface changes to black. Symptoms appear at the early developmental process where the fruit size is like a table tennis ball and it grows rapidly. BER is caused by various overlapping factors. The easiest way to reproduce BER artificially is to reduce the amount of Ca in the soil or culture solution. When Ca is eliminated from the culture solution in hydroponics, the fruit is slightly enlarged after fertilization and eventually develops BER. When Ca is severely reduced, the shoot apex will become necrotic, and the flower stalks will be damaged. When Ca is reduced to the extent that the shoot apex does not die, the fruit will exhibit BER.

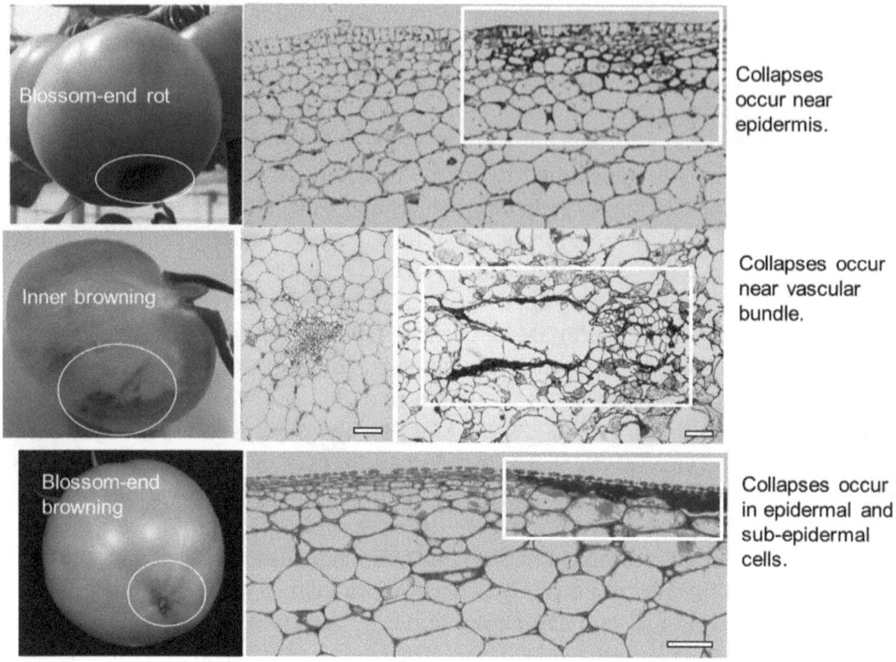

Fig. 6.4 Physiological disorders caused by Ca deficiency (bars: 100 μm)

Ca is an element that forms cells and is contained within cell walls and cell membranes; it also works as a second messenger in the cells. When Ca levels decrease, the cell membrane is destroyed, the cell wall cannot support the cell and it softens; the cell then collapses (Fig. 6.4). Once Ca is distributed to an organ, it is difficult to move to another organ. When Ca is deficient, it causes tip burn symptoms at the edge of the leaf, the shoot apex is terminated, and the end of the fruit may undergo necrosis. Ca is absorbed as Ca^{2+} ions from the roots and is transported to each tissue with water through a xylem. Ca hardly becomes deficient in leaves in which transpiration is active, but symptoms tend to appear on the shoot apex, the tip of the leaf during development, and the developing fruit where transpiration is not active. In addition, as these tissues require Ca for growth, if the amount of Ca required exceeds that supplied, cellular activity will be hindered and cells will die.

There have been many studies on BER (de Freitas et al. 2012; Ho and White 2005; Saure 2014). It was reported that a local deficiency of Ca will induce cell necrosis (Suzuki et al. 2003). There are various processes that occur before cell death in BER, and BER is caused by various conditions and overlapping environmental factors. BER is likely to occur under the following circumstances. Low Ca: The amount of Ca absorbed by the roots decreases (de Kreij 1996; Franco et al. 1999; Taylor et al. 2004). Strong light, high temperature, low humidity (high VPD): As transpiration from the leaves increases, Ca is preferentially transported to leaves and its transport to fruits decreases. At the same time, as fruit hypertrophy is promoted, Ca demand increases in the fruit (Adams and Holder 1992; de Kreij 1996). High humidity (low VPD): The amount of transpiration of the whole plant decreases and the amount of Ca transported to the fruit decreases (Barker 1990). High nitrogen: As the stem and leaves become overgrown and transpiration from the leaves increases, Ca is preferentially transported to the leaves and its transport to fruits decreases. At the same time, when the fruit enlargement is promoted, the demand of Ca increases in the fruit. High ammonium nitrogen: Ca absorption by the roots is inhibited (Taylor et al. 2004). High electrical conductivity (EC) and high salt concentration: Water absorption by the roots and Ca absorption are inhibited (Ho and White 2005). Low moisture: As the amount of water absorbed by the roots decreases, the amount of Ca transported to the fruit decreases (Pill and Lambeth 1980). High temperature/hypoxia: Nutrient absorption by the roots is inhibited and the amount of Ca transported to fruits decreases (Tachibana 1991).

Recent studies revealed factors responsible for BER as follows. It was suggested that a Ca^{2+} transport property was an essential factor for the lower incidence of BER when a tomato introgression line was used (Ikeda et al. 2017). The incidence of BER under salinity was reduced by the use of an organ-mineral fertilizer (Kataoka et al. 2017). It was suggested that the cultivar difference in the susceptibility to BER was likely to be attributable to the difference in the growth rate of young fruit, affecting water-soluble Ca in the distal part of tomato fruit (Vinh et al. 2018). The involvement of reactive oxygen species (ROS) is a major protagonist in the appearance of BER, and the ROS scavenging capacity in BER-resistant cultivars increased under salt stress conditions (Rached et al. 2018).

It is important to deal with the factors mentioned above appropriately in order to suppress the incidence of BER. To prevent a reduction in Ca absorption by the roots, an adequate amount of Ca should be applied, and high ammonia nitrogen should be avoided. In addition, drought stress and high-temperature/low oxygen condition should be avoided. Making active and vigorous roots is important as well.

It is also important to balance the growth of the stem and leaves, and the growth of fruits. If the fruit set ratio decreases due to high temperature and the temperature subsequently returns to optimal, the ratio of the amount of leaves to that of fruit will increase, likely resulting in excess plant vigor. Under such conditions of so-called overgrowth, Ca is preferentially transported to leaves rather than fruit, leading to BER. It is important to manage the sink–source balance at an appropriate level, so that plant overgrowth does not occur.

6.4.3 Leaf Marginal Necrosis and Yellow Shoulder Fruit Induced by Potassium Deficiency

Tomato's leaf marginal necrosis is a physiological disorder in which the tip (leaf edge) of the leaf turns yellow and eventually exhibits necrosis. If the tip of the leaf is only slightly yellowed, it is not a major problem. However, if the leaf yellowing is severe, it is likely that the photosynthetic rate has also decreased. In addition, the necrotic part becomes the primary infection source of disease (*Botrytis cinerea*).

Generally, leaf marginal necrosis is caused by potassium (K) deficiency in the leaves. The K content at the tip of the leaf tends to decrease in yellow leaves (Ueno et al. 2018). K is a highly mobile element in plants, and even if K deficiency occurs in the plant body, K easily moves from other parts of the plant and is supplied to the parts where it is most necessary. The necrosis of the tip of the old leaf is thought to be mainly due to K deficiency. Even when K is sufficient in the soil, the absorption of K decreases when high-humidity conditions continue due to cloudy or rainy weather. Decreased K absorption is commonly observed during sunny weather after rain even when the irrigation is sufficient. It is conceivable that low transpiration from the leaf becomes a trigger for K shortage.

Gene expression analyses of the basal and tip regions of leaflets revealed that the expression patterns of jasmonate-related genes were upregulated in the tomato leaflets with a low K concentration (Ueno et al. 2018). The gene expression of leaf senescence was detected only in the leaf tip region samples with the lowest K concentration. In addition, tomato plants cultivated with K-supplemented medium showed an increase in the K concentration, a decrease in the occurrence of leaf marginal necrosis, and downregulation of jasmonate-related gene expression in tip leaflets (Ueno et al. 2018). These results indicate that tomato leaf marginal necrosis occurs because of K starvation in the tip region of leaflets, leading to the activation of a jasmonate-induced signal for necrosis (Ueno et al. 2018).

According to a research report from Japan's Gifu Prefecture (where leaf marginal necrosis has become a serious problem in the production areas), leaf marginal necrosis in soil cultivation can be reduced by increased fertilization with potassium sulfate (Kumazaki et al. 2010). However, it was not possible to inhibit leaf marginal necrosis by soil application of an aqueous solution of potassium sulfate, or by applying slow-release potassium fertilizer in the summer cultivation. Foliar spraying with sulfuric acid reduces leaf marginal necrosis. An investigation of the frequency and application concentration of foliar application confirmed the beneficial effect of spraying twice a week, but a sufficient effect was not immediately observed. Leaf marginal necrosis was reduced by defoliation treatment, and this leaf defoliation treatment induced a high K concentration in the remaining leaves.

Yellow shoulder fruits appear easily in summer tomato cultivation (Suzuki et al. 2013). Yellow shoulder fruit occurs when the fruit surface is directly irradiated with strong sunlight during fruit development (Table 6.1). This can be demonstrated by covering the fruit (Fig. 6.5). The occurrence of this yellow shoulder fruit increases due to a K deficiency (Zhang et al. 2015a, b). It has been shown that by raising the concentrations of K and phosphoric acid, the absorption of K is promoted and the occurrence of this disorder is reduced in hydroponics (Zhang et al. 2019). By combining the method for promoting potassium absorption and the shading of fruits (Table 6.1, Fig. 6.5), the occurrence of yellow shoulder fruits under high temperature can be reduced.

Table 6.1 Ratio of yellow shoulder fruits with or without (control) covering fruits in three cropping

Planting date	Harvest date		Average air temperature	Average solar radiation	Percentage of yellow shoulder fruits		
					Control	With covering fruits	significance [a]
	First	Final	°C	MJm^{-2}	%	%	
2/19	4/17	5/25	20.6	16.7	53.3	27.1	*
3/17	5/13	6/30	21.3	18.4	66.2	42.8	*
6/8	7/13	8/10	27.1	15.8	21.1	15.7	ns

[a]*ns* indicates no significance, * indicates significance at 5% between percentage of yellow shoulder fruits in control and that with covering fruits by Student's *t*-test

Fig. 6.5 Fruit covered with fan-shaped aluminum foil and the fruit truss shaded with white sheet

6.5 Conclusions

Various problems have arisen in crop production due to global warming in recent years. Plant production in greenhouses is no exception, and it is necessary to take measures against abnormal weather, particularly against high temperatures in summer. Plant growing in high temperatures showed an increased incidence of various physiological disorders. It is possible to use air cooling by horsepower (HP), fog cooling, shading with new functional films, etc., which can lower the temperature. In the future, such equipment costs will decrease due to technological innovation. However, at present the costs are high. Therefore, it is important not only to lower the temperature but also to control the incidence of physiological disorders by economically controlling the fertilizer, growth, and environment. In the future, it is thought that various unexpected physiological disorders will occur due to global warming, and it is important to clarify the causes of these incidences and to develop low-cost countermeasures.

References

Adams P, Holder R (1992) Effects of humidity, Ca and salinity on the accumulation of dry matter and Ca by the leaves and fruit of tomato (*Lycopersicon esculentum*). J Hortic Sci 67:137–142

Aichi agricultural research center (2012) Development of high temperature control technology using mist. http://www.pref.aichi.jp/nososi/seika/singijutu/singijiyutu100.pdf (in Japanese)

Ainsworth EA, Ort DR (2010) How do we improve crop production in a warming world? Plant Physiol 154:526–530

Azzi L, Deluche C, Gévaudant F et al (2015) Fruit growth-related genes in tomato. J Exp Bot 66:1075–1086

Barker JC (1990) Effects of day and night humidity on yield and fruit quality of glasshouse tomatoes (*Lycopersicon esculentum* Mill.). J Hortic Sci 65:323–331

Bünger-Kibler S, Bangerth F (1983) Relationship between cell number, cell size and fruit size of seeded fruits of tomato (*Lycopersicon esculentum* Mill.), and those induced parthenocarpically by the application of plant growth regulators. Plant Growth Regul 1:143–154

de Freitas ST, Handa AK, Wu Q et al (2012) Role of pectin methylesterases in cellular calcium distribution and blossom-end rot development in tomato fruit. Plant J 71:824–835

de Kreij C (1996) Interactive effects of air humidity, calcium and phosphate on blossom-end rot, leaf deformation, production and nutrient contents of tomato. J Plant Nutr 19:361–377

Dorais M, Demers D-A, Papadopoulos AP et al (2004) Greenhouse tomato fruit cuticle cracking. Hortic Rev 30:163–184

Driedonks N, Rieu I, Vriezen WH (2016) Breeding for plant heat tolerance at vegetative and reproductive stages. Plant Reprod 29:67–79

Franco JA, Perez-Saura PJ, Fernandez JA et al (1999) Effect of two irrigation rates on yield, incidence of blossom-end rot, mineral content and free amino acid levels in tomato cultivated under drip irrigation using saline water. J Hortic Sci Biotechnol 74:430–435

Fuluoka N, Masuda D, Kanamori Y (2009) Effects of temperature around the fruit on sugar accumulation in watermelon (Citrullus lanatus (Thunb.) Matsum. & Nakai) during the latter half of fruit developmental period. J Jpn Soc Hortic Sci 78:97–102

Guichard S, Gary C, Leonardi C, Bertin N (2005) Analysis of growth and water relations of tomato fruits in relation to air vapor pressure deficit and plant fruit load. J Plant Growth Regul 24:201–213

Hiei K, Ito M, Ban Y, Oyabu T (2015) Effects of super-micro-fog spraying combined with shading on suppressing atmospheric temperature increase in greenhouse tomato cultivation under natural ventilation. Res Bull Aichi-ken Agric Res Cent 47:41–49 (In Japanese)

Ho LC, White P (2005) A cellular hypothesis for the induction of blossom-end rot in tomato fruit. Ann Bot 95:571–581

Ikeda H, Shibuya T, Nishiyama M et al (2017) Physiological mechanisms accounting for the lower incidence of blossom-end rot in tomato introgression line IL8-3 fruit. Hortic J 86:327–333

Ikeuchi M, Nimura M, Yamaguchi N et al (2012) Effect of super micro mist spraying on greenhouse environment and growth of pot plants in summer. Res Bull Aichi-ken Agric Res Cent 44:61–69 (In Japanese)

Iwasaki Y, Yoshida C, Aizawa M (2011) Effect of night cooling on yield and electricity consumption in summer to autumn culture of tomato. Hortic Res 10(Suppl.1):107 (In Japanese)

Kataoka K, Sugimoto K, Ohashi H et al (2017) Effect of organo-mineral fertilizer on tomato fruit production and incidence of blossom-end rot under salinity. Hortic J 86:357–364

Kawasaki Y, Ahn D-H (2015) Local cooling at night around flowering trusses improved yield of tomato grown under hot conditions. J SHITA 27:137–143 (In Japanese)

Kimura M, Fujitani S, Itimanda K (2012) Mitigation techniques on fruit cracking in tomato cultivation under rain shelter in summer and autumn I. Bull Oita Agric Forest Fish Res Cent 2:23–42 (In Japanese)

Kinomoto M, Matsumoto Y, Yoshida T (2013) Effect of humidity on yield and the biological information of tomato plants in the greenhouse. Bull Tochigi Agric Exp Stn 71:27–31 (In Japanese)

Kumazaki A, Kawakami M, Asano Y, Suzuki T (2010) Effects of fertilization method to suppress burned leaf tip of tomato. Hortic Res 9(Suppl.1):135 (In Japanese)

Matsuo S, Kikuchi K, Fukuda M et al (2012) Roles and regulation of cytokinins in tomato fruit development. J Exp Bot 63:5569–5579

Ohishi N, Moriya H (2011) Effect of the night environment control in middle summer season using heat pumps on plant and fruit growth of tomato cultivation. Hortic Res 10(Suppl.1):108 (In Japanese)

Ohishi N, Futamata N, Moriya H (2015) Effect of the night cooling and dehumidification in a high temperature season on fruit cracking of single-truss tomato by hydroponics. Hortic Res 14(Suppl.1):126 (In Japanese)

Ohta I, Takeda M (2014) Effects of UV-absorbing films on dispersal behavior of greenhouse whitefly, *Trialeurodes vaporariorum*. Jpn J Appl Entomol Zool 58:303–312 (In Japanese)

Paupière MJ, Mu¨ller F, Li H et al (2017) Untargeted metabolomic analysis of tomato pollen development and heat stress response. Plant Reprod 30:81–94

Pill WG, Lambeth VN (1980) Effects of soil water regime and nitrogen form on blossom-end rot, yield, water relations, and elemental composition of tomato. J Am Soc Hortic Sci 105(5):730–734

Rached M, Pierre B, Yves G et al (2018) Differences in blossom-end rot resistance in tomato cultivars is associated with total ascorbate rather than calcium concentration in the distal end part of fruits per se. Hortic J 87:372–381

Rieu I, Twell D, Firon N (2017) Pollen development at high temperature: from acclimation to collapse. Plant Physiol 173:1967–1976

Sakiyama H, Tsuchiya K, Udagawa Y (2004) Development of stable year-round production system in tomato and reduction of labor load by using transportable cultivation benches. 1.Outline of experimental facilities, working efficiency and reduction of labor load for thermal in high temperature environment of greenhouse. Bull Chiba-Ken Agric Res Cent 3:13–23 (In Japanese)

Sano O, Hikawa M, Imanishi S (2018) Reduction of radial fruit cracking by single spraying of forchlorfenuron (1-(2-chloro-4-pyridyl)-3-phenylurea) of fruit clusters in tomato production under rain shelter. Hortic Res 17:87–93 (In Japanese)

Sasaki H, Yano T, Yamasaki A (2005) Reduction of high temperature inhibition in tomato fruit set by plant growth regulators. JARQ 39:135–138

Saure MC (2014) Why calcium deficiency is not the cause of blossom-end rot in tomato and pepper fruit-a reappraisal. Sci Hortic 174:151–154

Suzuki T (2010) Studies on the cause of radial cracking and development of measures technology in tomato cultivation under rain shelter in cool upland. Bull Gifu Res Inst Agric Sci Hil Mount Areas 6:26–49

Suzuki K, Shono M, Egawa Y (2003) Localization of calcium in the pericarp cells of tomato fruit during the development of blossom-end rot. Protoplasma 222:149–156

Suzuki K, Kimura I, Kawasaki Y et al (2012) Morphological studies on change of peduncle after physical treatment in tomato. Hortic Res 11:569–575 (In Japanese)

Suzuki K, Sasaki H, Nagata M (2013) Causes and control of blotchy ripening disorder in tomato fruit. Natl Agric Food Res Org 12:81–88 (In Japanese)

Suzuki K, Suzuki D, Numajiri M, Matsushiro J, Yamane M, Kiriiwa Y (2019) Anatomical analysis of internally brown tomato fruit. Hortic J 88:263–269

Tachibana S (1991) Import of calcium by tomato fruit in relation to the day-night periodicity. Sci Hortic 45:235–243

Taylor MD, Locascio SJ, Alligood MR (2004) Blossom-end rot incidence of tomato as affected by irrigation quantity, calcium source, and reduced potassium. HortScience 39(5):1110–1115

Ueno H, Maeda T, Katsuyama N et al (2018) Cation measurements and gene expression analysis suggest tomato leaf marginal necrosis is caused by a jasmonate signal induced by k+ starvation in the tip region of leaflets. Hortic J 87:206–213

Uetani M, Fujitani S, Kimura M (2014) Mitigation techniques on fruit cracking in tomato cultivation under rain shelter in summer and autumn II. Bull Oita Agric Forest Fish Res Cent 4:11–25 (In Japanese)
Vinh TD, Yoshida Y, Ooyama M et al (2018) Comparative analysis on blossom-end rot incidence in two tomato cultivars in relation to calcium nutrition and fruit growth. Hortic J 87:97–105
Watanabe K, Shiwachi H, Iwahori S, Talahashi T (2006) Study on fruit cracking of tomato under greenhouse culture. J Agric Sci Tokyo Univ Agric 50:106–111 (In Japanese)
Yoshida Y, Hamamoto H, Fukunaga A et al (2008) Changes in the ascorbic acid content of shaded spinach after the removal of shading. Hortic Res 7:399–405 (In Japanese)
Zhang Y, Kiriiwa Y, Nukaya A (2015a) Influence of nutrient concentration and composition on the growth, uptake patterns of nutrient elements and fruit coloring disorder for tomatoes grown in extremely low-volume substrate. Hortic J 84:37–45
Zhang Y, Kiriiwa Y, Nukaya A (2015b) Effects of lower nitrogen concentration of nutrient solution combined with K supplementation and changing the concentration on growth, yield and yellow-shoulder disorder for tomatoes grown in extremely low-volume substrate. Hortic J 84:46–51
Zhang Y, Suzuki K, Liu H, Nukaya A et al (2019) Fruit yellow-shoulder disorder as related to mineral element uptake of tomatoes grown in high temperature. Sci Hortic 242:25–29

Chapter 7
Recent Improvements to Global Seasonal Crop Forecasting and Related Research

Toshichika Iizumi and Wonsik Kim

Abstract Recent changes in the frequency, intensity, and duration of weather and climate extremes and associated concerns for the stability of food supply and food price volatility in agricultural commodity markets underpin the importance of global crop monitoring and forecasting. Demands for monitoring and forecasting global food production remain high, even after the end of the food crises of 2010/2011. This article briefly overviews the progress during recent years and research outcomes regarding global seasonal crop forecasting achieved by National Agriculture and Food Research Organization (NARO) and collaborators. The research topics include global seasonal crop forecasting (Sect. 7.2.1), climate oscillations and global yields (Sect. 7.2.2), climate change and yield variability (Sect. 7.2.3), and global crop datasets (Sect. 7.2.4). We also discuss issues that need to be addressed by future research to further strengthen existing global crop monitoring and forecasting systems in the face of unprecedented extreme weather episodes in a warmer climate.

7.1 Introduction

Globalization of the economy has changed the food supply chain worldwide. Consumers in many countries have increasingly relied on food imports. Therefore, national governments and commercial entities in import-dependent countries pay close attention to food production and export prices in major food-exporting countries as well as their own domestic production. Demand for monitoring and forecasting global food production sharply increased in the face of the food crises of

T. Iizumi (✉) · W. Kim
Institute for Agro-Environmental Sciences, National Agriculture and Food Research Organization (NARO), Tsukuba, Ibaraki, Japan
e-mail: iizumit@affrc.go.jp

T. Iizumi et al. (eds.), *Adaptation to Climate Change in Agriculture*,
https://doi.org/10.1007/978-981-13-9235-1_7

2008 and 2010/2011. Although food prices in commodity markets have gradually declined since the end of the latter crisis, food prices remain high compared to those in the early 2000s (FAO 2018b).

Seasonal climate variability and extremes are main drivers of year-to-year variations in crop yields and sometimes lead to food price spikes, which directly worsen the consumers' access to affordable food especially in developing countries. A global analysis estimates that approximately one-third of global yield variability is explained by variation in seasonal climate conditions (Ray et al. 2015). More specifically, the drought and heat wave of 2010/2011 in Russia (Welton 2011), the 2012 drought in the United States of America (Rippey 2015), and the heat wave over Asia in 2016 (Imada et al. 2018) constitute recent examples of climate extremes affecting crop production. The Food and Agriculture Organization (FAO) of the United Nations reports that climate extremes, especially droughts in food insecure regions, and conflicts contribute to the recent rise in hunger (FAO 2018a).

Consequently, commodity traders and food security experts demand global agricultural monitoring and forecasting to help better respond to seasonal-climate-induced food production and price shocks. At the same time, these technologies help improve the ability of societies to adapt to climate change (IPCC 2014; Ceglar et al. 2018; Iizumi et al. 2018c). Two initiatives, the Agricultural Market Information System (AMIS) (FAO 2016a) and the Group of Earth Observations (GEO) Global Agricultural Monitoring (GEOGLAM), have operated since their launch at the 2011 Group of Twenty (G20) Cannes summit (G20 2011). GEOGLAM utilizes the Earth Observation Satellite (EOS) (Whitcraft et al. 2015) to provide information on crop growth and associated early warning derived from satellite remote sensing for AMIS (AMIS 2019). However, crop forecasting (or yield prediction) has not yet been included in these initiatives at this moment of writing. Therefore, research on crop monitoring and forecasting should contribute to strengthen the functions of these existing systems.

This chapter aims to report recent progress in global seasonal crop forecasting and related research conducted by NARO and collaborators to give readers a quick overview of these topics. The article includes global seasonal crop forecasting (Sect. 7.2.1), climate oscillations and global yields (Sect. 7.2.2), climate change and yield variability (Sect. 7.2.3), and global crop datasets (Sect. 7.2.4). We also discuss remaining issues in global seasonal crop forecasting that need to be addressed by future research.

7.2 Recent Progress in Global Seasonal Crop Forecasting and Related Research

Global seasonal crop forecasting and related key research topics are illustrated in Fig. 7.1. Seasonal crop forecasting is an emerging application of scientific understanding on climate–yield relationships to enable societies (food agencies and agro-businesses in particular) coping with risk in agricultural production and trade

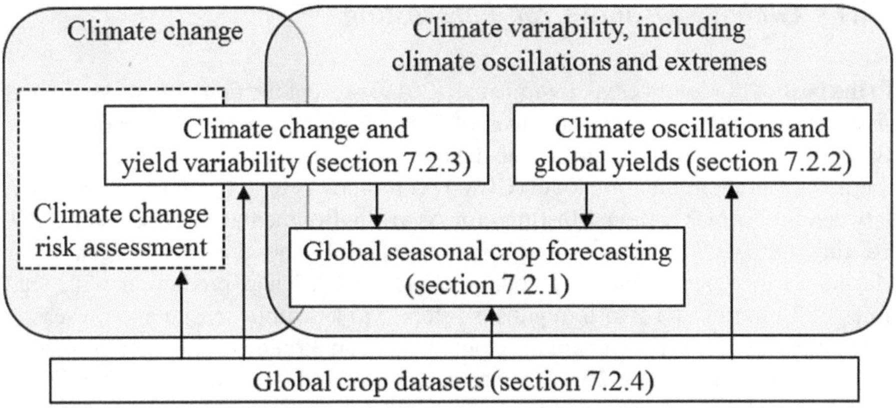

Fig. 7.1 Schematic illustrating the relationship between global seasonal crop forecasting and related research topics

associated with climate variability and change. The global crop datasets are a basis for climate–yield relationship analysis at regional to global scales. The development of a new global crop dataset is motivated to improve the functionality of global seasonal crop forecasting and climate change risk assessment in crop production systems. Climate oscillations, such as El Niño Southern Oscillation (ENSO), and their impacts on crop yields are a topic in the field on seasonal-to-interannual-scale climate–yield relationship analysis that have direct implications in improving seasonal crop forecasting and early warning over the large spatial domain. An assessment of climate change impacts on yield can be deemed as a decadal-to-century-scale climate–yield relationship analysis. Because evidence indicates that observed climate change has already led to changes in intensity, frequency, and duration of weather and climate extremes which decrease the quantity and quality of agricultural products, understanding of yield variability change and the relative contributions of climatic and nonclimatic drivers to the change is key to ensure the applicability of seasonal crop forecasting methods to a warmer climate in coming decades. Hence, both of short-term climate oscillation and long-term climate change are important and need to be considered in seasonal crop forecasting. Climate change risk assessment is outside the scope of this article (recent progress related to this topic achieved by NARO and collaborators is described in Iizumi (2019)).

In the following sections, we start from global seasonal crop forecasting (Sect. 7.2.1), of which operationalization is our goal in coming years. Then climate oscillations and their impacts on global yields (Sect. 7.2.2), which may offer a new yield prediction method, are presented. The linkage between climate change and yield variability (Sect. 7.2.3) which falls within the category of climate change risk assessment but has implications to improve global seasonal crop forecasting is elaborated. Lastly, global crop datasets (Sect. 7.2.4), which are a base of any climate–yield relationship analysis and their development requires basic research, are described.

7.2.1 Global Seasonal Crop Forecasting

To the best of our knowledge, Iizumi et al. (2013) present the first global assessment of yield prediction skill at lead times of 3 and 5 months before harvesting using nine-member ensemble mean climate data derived from a single ocean–atmosphere coupled general circulation model (GCM). This study is based on the joint research between the former National Institute for Agro-Environmental Sciences (currently, Institute for Agro-Environmental Sciences, NARO) and the Japan Agency for Marine-Earth Science and Technology (JAMSTEC). Yield prediction with lead times of 3 months and 5-to-6 months is referred to as within-season and preseason prediction, respectively. The monthly outputs of 2 m air temperature and root-zone soil moisture data for the 25-year period (1982–2006) derived from SINTEX-F1 GCM (Luo et al. 2005) are used to predict yield anomalies.

The key findings of Iizumi et al. (2013) include the following: moderate-to-remarkable yield anomalies (5% or larger deviation in year t from normal yield, defined as the average yield for the previous 3 years (t-3 to t-1)) are reliably predicted 3 months before harvest for rice and wheat, but not for maize and soybean; on a global basis, growing season temperature more largely contributes to rice and wheat yield variability than growing season soil moisture, whereas soil moisture is more important than temperature in explaining maize and soybean yield variability; and within-season and preseason predictions of smaller yield anomalies (<5%) are less reliable than those of moderate-to-remarkable yield anomalies because of weaker climate signals.

While scientifically novel, Iizumi et al. (2013) have some limitations. First, only one-fifth (15–19%) of the global harvested area has reliable yield predictability with varying extent of area by crop (the area produces 15–23% of the global production), which is insufficient for predicting national yield anomalies, which are of more direct interest for commodity traders and food security experts than local yield anomalies. Additional skilled areas are necessary to predict national yield anomalies. Second, maize and soybean are important agricultural commodities traded worldwide and hence yield prediction information for these crops is desired, especially for consumers in import-dependent countries. Iizumi et al. (2016) suggest that global seasonal crop forecasting would benefit more from additional skillful seasonal climate forecasts than from additional accurate climate–yield modeling in terms of increasing the extent of skilled areas. If so, use of multimodel ensemble (MME) seasonal climate forecast data is a promising approach for improving yield prediction skills at the global scale.

Iizumi et al. (2018c) present the latest global assessment of yield prediction skill using an improved prediction method. By using MME monthly temperature and precipitation data derived from five different GCMs, Iizumi et al. (2018c) show that the within-season yield prediction of over 25–38% of the global harvested area is reliably feasible, with varying extent of area by crop. This updated assessment, which is an outcome from the joint research project of NARO and the Asia-Pacific Economic Cooperation (APEC) Climate Center (APCC), is based on data for the 1983–2010 period at the 1.125°-grid-cell level and for the 1983–2015 period at the national level. Importantly, the reliable yield prediction of maize and soybean becomes feasible; for rice and wheat, the extent of skillful area presented in Iizumi et al. (2018c) increases from that shown in earlier work (Iizumi et al. 2013); and within-season and preseason predictions of national yield variability for some country–crop combinations, including major food exporters, such as wheat in Australia and maize and soybean in Argentina, are found to be skillful, due to the increase in the skilled area. Based on the yield prediction method tested in Iizumi et al. (2018c), APCC and NARO plan to conduct a test operation of global seasonal crop forecasting service in coming growing seasons (2019/2020).

7.2.2 *Climate Oscillations and Global Yields*

Prediction skills for large-scale oceanic oscillations, such as ENSO, are much higher than those for seasonal temperature and moisture at local scales. Because weather centers and research institutes worldwide have long performed ENSO forecasting, empirical relationships between local yield anomalies and the ENSO phase are useful for improving preparedness toward anticipated production and price shocks associated with warmer (El Niño) and cooler (La Niña) phases of ENSO (e.g., Ubilava (2017)). Iizumi et al. (2014a) depict global patterns of average ENSO impacts on yields of major crops from 1982 to 2006 at the 1.125° resolution. Importantly, significant negative or positive yield impacts due to ENSO are found even for higher latitudes where seasonal climate forecast skills are generally lower than those in the low latitudes. The geographic patterns of ENSO–yield relationships revealed in Iizumi et al. (2014a) improved our knowledge of the ENSO impacts on global agriculture, which is a notable update that was mainly based on the findings of Rosenzweig and Hillel (2008). Food agencies have started using improved knowledge, including that mentioned above, to improve the preparedness to ENSO impacts (FAO 2014) and provide early warnings (Iizumi 2014, FAO 2016b).

However, Iizumi et al. (2014a) do not tell us if a stronger El Niño (or La Niña) event influences yields more severely than a weaker event, which is a known weakness of Iizumi et al. (2014a). Additionally, the relative importance of different climate oscillations in explaining yield anomalies may vary by location and season; however, in many cases, ENSO appears to be the most dominant climate mode at the seasonal to multiyear scales.

The hypothesis mentioned above is examined by some studies. Heino et al. (2018) compare three different climate oscillations – ENSO, Indian Ocean Dipole (IOD), and North Atlantic Oscillation (NAO) – and address the most influential oscillation type for crop production by region. However, average yields across multiple crop types simulated by a global gridded crop model are used in Heino et al. (2018), and therefore, crop-specific information is necessary to be informative because different crop types have different economic values. Yuan and Yamagata (2015) present an analysis examining the Australian wheat yield responses to IOD and two types of ENSO (canonical ENSO and ENSO Modoki) and demonstrate that

IOD is relatively more important than the two ENSO types in explaining the variations in the yields in Western Australia, which is an area adjacent to the Indian Ocean. In contrast, canonical ENSO and ENSO Modoki are relatively more powerful in explaining wheat yield variability in Eastern Australia, which is an area adjacent to the Pacific Ocean (Yuan and Yamagata 2015). Ceglar et al. (2017) report statistically significant relationships between national yield anomalies of winter wheat and grain maize and four atmospheric oscillation types (NAO, Eastern Atlantic, Scandinavian, and Eastern Atlantic–Western Russia patterns) in Europe. Anderson et al. (2017) reveal that the yield impacts of ENSO in Brazil have changed with increased annual number of harvests.

Yield prediction skills based on forecasted climate oscillation indices are already tested for Australian wheat (Yuan and Yamagata 2015) and European summer grain maize (Ceglar et al. 2018). A consistent comparison between two different yield prediction methods, one based on temperature and moisture anomalies (e.g., Iizumi et al. 2013, 2018c) and the other based on climate oscillation indices (Yuan and Yamagata 2015; Ceglar et al. 2018), is awaited to seek a better method for global seasonal crop forecasting.

7.2.3 Climate Change and Yield Variability

Recent changes in the frequency, intensity, and duration of weather and climate extremes are evident (Alexander 2016). In addition to change in mean climate, the changes in weather and climate extremes are important because these raise concerns for the stability of food supply and food price volatility in agricultural commodity markets. Osborne and Wheeler (2013) examine the historical changes in national yield variability in the top crop-producing countries for the last half century (1961–2010) and find that, in 7 out of 30 country–crop combinations, changes in yield variability are related to changes in climate variability. Therefore, whether the recent changes in daily temperature and precipitation extremes have a measurable influence on yield variability needs to be addressed to deepen our understanding of climate change impacts on yield variability and improve global seasonal crop forecasting.

Iizumi and Ramankutty (2016) reveal that a significant increase in yield variability over the last three decades (1981–2010) is found in 9–22% of the global harvested area, with varying extent of area by crop, although a decrease in yield variability is the main trend worldwide across crops (19–33%). More importantly, approximately 20% of the detected yield variability change is attributed to climate change and the associated shift in modeled crop phenology. Among others, more frequent high temperatures exceeding the optimal temperature range suitable for yield formation are the climatic factor that best explains the detected yield variability change. Note however that nonclimatic factors (e.g., input price volatility, social unrest, and conflicts) play a larger role than climatic factors to lead to the yield variability change.

The uncertainty of the estimated extent of areas with increased (decreased) yield variability over the period from 1981 to 2008 is evaluated using two different global yield datasets at four spatial resolutions (0.083°, 0.5°, 1°, and 2°) (Iizumi et al. 2018a). One of the datasets (Iizumi et al. 2014b) is largely based on satellite remote sensing and the other (Ray et al. 2012) is solely compiled from agricultural census. Iizumi et al. (2018a) represent the first uncertainty assessment of historical changes in yield variability based on two different yield datasets, as earlier assessments were driven by the single best available yield dataset. The detected yield variability changes in rice and wheat are sensitive to the choice of dataset and resolution to some degree, but they are relatively robust for maize and soybean (Iizumi et al. 2018a). Although quantitative uncertainty in estimated extent of area persists, the intercomparison results corroborate the findings of Iizumi and Ramankutty (2016) that yields in some regions of the world have become more unstable; for wheat and rice, the extent of area with increased yield variability is comparable to or larger than that with decreased yield variability. These findings reinforce the need for global yield monitoring and forecasting to increase the ability of societies to better respond to food production and price shocks associated with climate extremes under both present-day and future climates.

The main climatic disaster types damaging crop production vary by region and season, although droughts are frequently noted as the most severe disaster for global agricultural production (Lesk et al. 2016). Even recently, droughts and associated vapor pressure deficits remain key climate drivers of maize yield variability in the central United States (Lobell et al. 2014). Thus, Kim et al. (2019) estimate detailed geographic patterns of drought-induced yield loss and economic damage. However, heavy precipitation with high temperatures is gradually being recognized as an emerging climatic factor for wheat yield loss (Zampieri et al. 2017). Because Ben-Ari et al. (2018) point out that the European operational crop forecasting system failed to predict the extreme wheat yield loss of 2016 in France due to an unprecedentedly hot autumn and wet spring; the understanding of historical climate change and its impacts on yield variability needs be improved, which is key to updating the yield prediction methods used in operational crop forecasting. In addition to climate change, nonclimatic factors, such as agronomic technology and management (in particular, irrigated area extent; Kucharik and Ramankutty 2005; Hawkins et al. 2013), crop pests and pathogens (Bebber et al. 2013, 2014), and their historical change are important to lead to a more skillful yield prediction.

7.2.4 Global Crop Datasets

Crop production is a function of the number of harvests per year, the area harvested, and the production per unit area (yield) (Iizumi and Ramankutty 2015). The availability of global datasets on crop calendars, harvested area, and yield are therefore essential as a basis for global seasonal crop forecasting and climate–yield relationship analysis. Daily to sub-daily weather data are also important, but in general, crop data are often hard to access compared to weather data (Iizumi et al. 2018b).

High-resolution gridded yield datasets are desired for a better understanding of climate–yield relationships because country-level yield and climate data are often too smoothed to detect the yield impacts associated with weather and climate extremes (a notable exception is Lesk et al. (2016)). A global dataset of yield for different crops at subnational resolution has only recently become available. Ray et al. (2012) are the first to compile a spatially explicit global dataset of major crops for the 1961–2008 period with a spatial resolution of 5-minute arc (though the exact resolution is national or subnational scale because a "paint-by-number" approach is used to compile the dataset). The yield dataset of Ray et al. (2012) is based on over 2.5 million agricultural statistics, collected for over 13,500 political units across the world. Ray et al. (2012) collect harvested area data as well as yield data because harvested areas are important when aggregating grid-cell yields into a country average yield. Unfortunately, however, the dataset of Ray et al. (2012) is not fully open for research communities at the moment of writing.

Another yield dataset for major crops is developed in Iizumi et al. (2014b) (referred to as the Global Dataset of Historical Yields or GDHY). The GDHY dataset initially covered the 1982–2006 period (the version 1) and was extended to cover the 1981–2011 period (the version 1.1), but both versions have the common grid size of 1.125°. The latest GDHY dataset (version 1.2) improves its resolution (0.5°), while the time coverage is the same with earlier version (1981–2011). GDHY dataset version 1.2 is available via Data Integration and Analysis System (DIAS, https://doi.org/10.20783/DIAS.528). All GDHY dataset versions are a hybrid of FAO national yield statistics and satellite-derived crop-specific vegetation index (i.e., net primary production). Therefore, the grid-cell yields available in the GDHY datasets are model estimates rather than observed yields. A more detailed description of GDHY dataset versions is available in Iizumi et al. (2018a). An update to cover the period up to 2016 at the 0.5° resolution is now ongoing.

Currently, global crop calendars are available from the Center for Sustainability and the Global Environment, University of Wisconsin-Madison (SAGE, Sacks et al. 2010) and the global dataset of monthly irrigated and rainfed crop areas around the year 2000 (MIRCA2000, Portmann et al. 2010). Both datasets are based on governmental crop progress reports from the 1990s to the early 2000s. SAGE dataset reports the planting and harvesting dates of 19 major crops and distinguishes the growing season (e.g., winter and spring wheat). The MIRCA2000 dataset reports the planting and harvesting months of 26 crops and differentiates irrigated and rainfed conditions. No distinction is available between growing seasons in MIRCA2000. Although the resolution of these datasets is 5-minute arc, a single data value is often assigned for all grid cells within a country because a "paint-by-number" approach is used. The MIRCA2000 dataset entirely covers the global cropland in 2000, whereas the spatial coverage of the SAGE dataset varies by crop. Although these datasets represent the best global crop calendar information currently available, a caution is necessary because, for some cases, discrepancies in crop calendars across data sources persist (Waha et al. 2012; Mathison et al. 2017; Iizumi et al. 2019). For rice, a more detailed global dataset has been recently developed (Laborte et al.

2017), and it is expected that global crop calendar modeling (Waha et al. 2012; Iizumi et al. 2019) plays a role in estimating detailed geographic patterns of crop calendars, which could be used as an input for yield prediction methods.

7.3 Knowledge Gaps in Global Seasonal Crop Forecasting

Some issues in global seasonal crop forecasting remain, and these issues need to be addressed by future research. Such issues include understanding of drivers of yield variability (Sect. 7.3.1), uncertainties of global crop datasets (Sect. 7.3.2), characterizing yield prediction methods (Sect. 7.3.3), and nonclimatic factors in yield prediction (Sect. 7.3.4). The following sections describe more details of the individual issues.

7.3.1 Drivers of Yield Variability

First, an improved understanding of the characteristics of yield prediction skill is required. Although Iizumi et al. (2018c) increased the extent of area with certain yield predictability to about one-third of the global harvested area, yield prediction skill in the remaining portion of the world's cropping area is currently low. It is therefore a mandate to know what climatic and/or nonclimatic factors cause yield anomalies in that area. Studies addressing this issue, such as Hawkins et al. (2013), Ray et al. (2015), Schauberger et al. (2017), and Ben-Ari et al. (2018), are useful to deepen our understanding on this issue.

In addition, geographic and seasonal patterns of yield prediction skill need to be characterized. It is known that the yield prediction skill is a function of the reliability of seasonal climate forecast and the strength of climate–yield relationship (Iizumi et al. 2016). Wilks and Godfrey (2002) characterized the climate prediction skills as follows: Climate forecasts in the low latitudes are more reliable than in the mid-to-high latitudes; the prediction skills for climate forecasts in northern hemispheric winter are higher than those in summer; and temperature forecasts are more reliable than moisture forecasts. However, crop calendars, yield-determining climatic factor, and agronomic management are different across crops and regions. This makes characterizing yield prediction skills at the global scale complicated.

7.3.2 Uncertainties of Global Crop Datasets

Second, an improved understanding of the uncertainties of global crop datasets is required. Multiple datasets, such as crop calendars, harvested area, and yields, are available for essential crop variables in global seasonal crop forecasting. For instance, it is known that the use of different harvested area datasets leads to substantial differences in calculated global mean yields (Porwollik et al. 2017). An intercomparison of multiple crop datasets enables us to characterize the uncertainties of predicted yields and the production outlooks associated with the use of different crop datasets. Although no study is available in the field of global seasonal crop forecasting, such analysis is found in other disciplines (e.g., See et al. (2015), Anderson et al. (2015), and Iizumi et al. (2018a)).

7.3.3 Characterizing Yield Prediction Methods

Third, advantages and disadvantages of different yield prediction methods need to be characterized. For instance, statistical yield models have much higher applicability than ecophysiological process-based crop models. Many attempts have been conducted to apply seasonal climate forecast data to process-based crop modeling at farm field scales to inform farmers about possible adjustment of agronomic practices in response to predicted seasonal climate variability (Sivakumar and Hansen 2007; Hayashi et al. 2018; van der Velde and Nisini 2019; Hayashi et al. 2019). However, process-based models have high input requirements, including management information that is in general hard to access. Hence, it is often difficult to use field-scale process-based crop models for operational crop forecasting with a few exceptions (for instance, the WOrld FOod Studies(WOFOST) model is used in the operational yield forecasting system in Europe (FAO 2016a; van der Velde and Nisini 2019)). Combining a field-scale process-based model with satellite remote sensing is a possible solution for some cases (Maki et al. 2017; Azzari et al. 2017). Global gridded crop models that are recently emerging in the field on climate change risk assessment and have capability to simulate crop yields over the globe from one grid cell to another (Müller et al. 2017) may be another candidate (FAO 2017). Yield prediction methods solely based on satellite remote sensing are also available (Johnson 2014; López-Lozano et al. 2015; Johnson 2016). A consistent comparison in yield prediction skill between yield prediction methods is expected to be useful for researchers when selecting appropriate yield prediction methods for global seasonal crop forecasting.

7.3.4 Nonclimatic Factors in Yield Prediction

Lastly, seasonal climate variability is a predominant driver of year-to-year yield variability, but it is not a sole determinant of yield in absolute terms (yield level). Because estimates of yield level are essential for deriving production outlooks, current statistical yield models that only predict yield anomalies (Iizumi et al. 2013, 2018c; Schauberger et al. 2017), relative to normal yield, are insufficient for fulfilling user requirements and thus need to be overhauled. However, currently we know little about which nonclimatic variables can be used to predict yield level. A better understanding of the relative contributions of nonclimatic factors, including technology, management, policy as well as climate to historical yield growth and variability, is necessary to tackle this issue. A good example in this research topic can be seen in Bernhard et al. (2018).

7.4 Conclusions

Seasonal crop forecasting based on seasonal climate forecast data provides predictions of yield but not production. By contrast, satellite remote sensing used for global agricultural monitoring is a powerful method of estimating cultivated or harvested area; however, yield predictions solely based on satellite remote sensing are still challenging. To understand the relative advantages and disadvantages of individual technologies, it is necessary to combine seasonal crop forecasting and satellite-based agricultural monitoring to enable possible variations in crop production to be predicted. The recent research outcomes in the field of global seasonal crop forecasting achieved by NARO and collaborators are expected to be useful to further strengthen existing global agricultural monitoring systems. Global seasonal crop forecasting has largely relied on underlying understanding of climate–yield relationships and global crop datasets. Research on these topics is needed to increase the capacity of societies to cope with short- and long-range climate risk in agricultural production.

Acknowledgments T.I. was partly supported by the Grant-in-Aid for Scientific Research (B, 16KT0036 and 18H02317; and C, 17 K07984) of the Japan Society for the Promotion of Science. T.I. and W.K. were partly supported by the Environment Research and Technology Development Fund (S-14) of the Environmental Restoration and Conservation Agency and the Joint Research Program of Arid Land Research Center, Tottori University (30F2001).

References

Alexander LV (2016) Global observed long-term changes in temperature and precipitation extremes: a review of progress and limitations in IPCC assessments and beyond. Weather Clim Extremes 11:4–16

AMIS (2019) Market Monitor. AMIS, Rome. http://www.amis-outlook.org/amis-monitoring. Accessed 24 Jan 2019

Anderson W, You L, Wood S et al (2015) A comparative analysis of global cropping systems models. Glob Ecol Biogeogr 24:180–191

Anderson W, Seager R, Baethgen W et al (2017) Crop production variability in North and South America forced by life-cycles of the El Niño Southern Oscillation. Agric For Meteorol 239:151–165

Azzari G, Jain M, Lobell DB (2017) Towards fine resolution global maps of crop yields: testing multiple methods and satellites in three countries. Remote Sens Environ 202:129–141

Bebber DP, Ramotowski MAT, Gurr SJ (2013) Crop pests and pathogens move polewards in a warming world. Nat Clim Chang 3:985–988

Bebber DP, Holmes T, Smith D et al (2014) Economic and physical determinants of the global distributions of crop pests and pathogens. New Phytol 202:901–910

Ben-Ari T, Boé J, Ciais P et al (2018) Causes and implications of the unforeseen 2016 extreme yield loss in the breadbasket of France. Nat Commun 9:1627. https://doi.org/10.1038/s41467-018-04087-x

Bernhard S, Ben-Ari T, Makowski D et al (2018) Yield trends, variability and stagnation analysis of major crops in France over more than a century. Sci Rep 8:16865. https://doi.org/10.1038/s41598-018-35351-1

Ceglar A, Turco M, Toreti A et al (2017) Linking crop yield anomalies to large-scale atmospheric circulation in Europe. Agric For Meteorol 240–241:35–45

Ceglar A, Toreti A, Prodhomme C et al (2018) Land-surface initialisation improves seasonal climate prediction skill for maize yield forecast. Sci Rep 8:1322. https://doi.org/10.1038/s41598-018-19586-6

FAO (2014) Understanding the drought impact of El Niño on the global agricultural areas: an assessment using FAO's Agricultural Stress Index (ASI). FAO, Rome. http://www.fao.org/3/a-i4251e.pdf. Accessed 18 Nov 2018

FAO (2016a) Crop yield forecasting: methodological and institutional aspects current practices from selected countries (Belgium, China, Morocco, South Africa, USA) with a focus on AMIS crops (maize, rice, soybeans and wheat). FAO, Rome. http://gsars.org/wp-content/uploads/2016/03/AMIS_CYF-Methodological-and-Institutional-Aspects_0303-web.pdf. Accessed 18 Nov 2018

FAO (2016b) 2015–2016 El Niño – early action and response for agriculture, food security and nutrition. FAO, Rome, http://www.fao.org/emergencies/resources/documents/resources-detail/en/c/340660. Accessed 18 Nov 2018

FAO (2017) Recent practices and advances for AMIS crop yield forecasting at farm and parcel level: a review. FAO, Rome. http://www.fao.org/3/a-i7339e.pdf. Accessed 18 Nov 2018

FAO (2018a) The state of food security and nutrition in the world 2018. Building climate resilience for food security and nutrition. FAO, Rome. https://reliefweb.int/sites/reliefweb.int/files/resources/English___The_State_of_Food_Security_and_Nutrition_in_the_World_2018_-_Full_Report.pdf. Accessed 26 Nov 2018

FAO (2018b) World Food Situation. FAO, Rome. http://www.fao.org/worldfoodsituation/foodpricesindex/en/. Accessed 22 Nov 2018

G20 (2011) Cannes Summit Final Declaration "Building Our Common Future: Renewed Collective Action for the Benefit of All". G20, Cannes. http://www.oecd.org/g20/summits/cannes/Cannes%20Declaration%204%20November%202011.pdf. Accessed 22 Nov 2018

Hawkins E, Fricker TE, Challinor AJ et al (2013) Increasing influence of heat stress on French maize yields from the 1960s to the 2030s. Glob Chang Biol 19:937–947

Hayashi K, Llorca L, Rustini S et al (2018) Reducing vulnerability of rainfed agriculture through seasonal climate predictions: a case study on the rainfed rice production in Southeast Asia. Agric Syst 162:66–76

Hayashi K, Llorca L, Bugayong I (2019) Decision support system for rainfed rice farming in Southeast Asia. In: Iizumi T, Hirata R, Matsuda R (eds) Adaptation to climate change in agriculture. Springer, Singapore (in this volume)

Heino M, Puma MJ, Ward PJ et al (2018) Two-thirds of global cropland area impacted by climate oscillations. Nat Commun 9:1257. https://doi.org/10.1038/s41467-017-02071-5

Iizumi T (2014) Outlook of global yield impacts due to El Niño 2014. In: Ministry of Agriculture, Forestry and Fisheries of Japan (MAFF). Oversea food demand and supply report (monthly report) 2014 July. MAFF, Tokyo. http://www.maff.go.jp/j/zyukyu/jki/j_rep/monthly/201407/pdf/21_monthly_topics-1.pdf. Accessed 25 Nov 2018

Iizumi T (2019) Improvements in climate change risk assessment for global crop production systems. In: Shirato Y, Hasebe A (eds) Climate smart agriculture for the small-scale farers in the Asian and Pacific region. NARO, Tsukuba and Food and Fertilizer Technology Center (FFTC)

for the Asian and Pacific Region, Taipei, pp 191–200. http://www.naro.affrc.go.jp/publicity_report/publication/files/fftc-marco_book2019.pdf
Iizumi T, Ramankutty N (2015) How do weather and climate influence cropping area and intensity? Glob Food Sec 4:46–50
Iizumi T, Ramankutty N (2016) Changes in yield variability of major crops for 1981–2010 explained by climate change. Environ Res Lett 11:034003. https://doi.org/10.1088/1748-9326/11/3/034003
Iizumi T, Sakuma H, Yokozawa M et al (2013) Prediction of seasonal climate-induced variations in global food production. Nat Clim Chang 3:904–908
Iizumi T, Luo J-J, Challinor AJ et al (2014a) Impacts of El Niño Southern Oscillation on the global yields of major crops. Nat Commun 5:3712. https://doi.org/10.1038/ncomms4712
Iizumi T, Yokozawa M, Sakurai G et al (2014b) Historical changes in global yields: major cereal and legume crops from 1982 to 2006. Glob Ecol Biogeogr 23:346–357
Iizumi T, Sakuma H, Yokozawa M et al (2016) Characterizing the reliability of global crop prediction based on seasonal climate forecasts. In: Behera SK, Yamagata T (eds) Indo-Pacific climate variability and predictability. World Scientific Publisher, Singapore, pp 281–304
Iizumi T, Kotoku M, Kim W et al (2018a) Uncertainties of potentials and recent changes in global yields of major crops resulting from census- and satellite-based yield datasets at multiple resolutions. PLoS One 13(9):e0203809. https://doi.org/10.1371/journal.pone.0203809
Iizumi T, Masutomi Y, Takimoto T et al (2018b) Emerging research topics in agricultural meteorology and assessment of climate change adaptation. J Agric Meteorol 74:54–59
Iizumi T, Shin Y, Kim W et al (2018c) Global crop yield forecasting using seasonal climate information from a multi-model ensemble. Clim Serv 11:13–23
Iizumi T, Kim W, Nishimori M (2019) Modeling the global sowing and harvesting windows of major crops around the year 2000. J Adv Model Earth Syst 11. https://doi.org/10.1029/2018MS001477
Imada Y, Shiogama H, Takahashi C et al (2018) Climate change increased the likelihood of the 2016 heat extremes in Asia. Bull Am Meteorol Soc 99:S97–S101
IPCC (2014) Summary for policymakers. In: Field CB, Barros VR, Dokken DJ et al (eds) Climate change 2014: impacts, adaptation, and vulnerability. Part A: global and sectoral aspects. IPCC, Cambridge, pp 1–32
Johnson DM (2014) An assessment of pre- and within-season remotely sensed variables for forecasting corn and soybean yields in the United States. Remote Sens Environ 141:116–128
Johnson DM (2016) A comprehensive assessment of the correlations between field crop yields and commonly used MODIS products. Int J Appl Earth Obs Geoinf 52:65–81
Kim W, Iizumi T, Nishimori M (2019) Global patterns of crop production losses associated with droughts from 1984 to 2009. J Appl Meteorol Climatol 58:1233–1244
Kucharik CJ, Ramankutty N (2005) Trends and variability in U.S. corn yields over the twentieth century. Earth Interact 9:1–29
Laborte AG, Gutierrez MA, Balanza JG et al (2017) RiceAtlas, a spatial database of global rice calendars and production. Sci Data 4:170074. https://doi.org/10.1038/sdata.2017.74
Lesk C, Rowhani P, Ramankutty N (2016) Influence of extreme weather disasters on global crop production. Nature 529:84–87
Lobell DB, Roberts MJ, Schlenker W et al (2014) Greater sensitivity to drought accompanies maize yield increase in the U.S. Midwest. Science 344:516–519
López-Lozano R, Duveiller G, Seguini L et al (2015) Towards regional grain yield forecasting with 1km-resolution EO biophysical products: strengths and limitations at pan-European level. Agric For Meteorol 206:12–32
Luo J-J, Masson S, Behera S et al (2005) Seasonal climate predictability in a coupled OAGCM using a different approach for ensemble forecasts. J Clim 18:4474–4497
Maki M, Sekiguchi K, Homma K et al (2017) Estimation of rice yield by SIMRIW-RS, a model that integrates remote sensing data into a crop growth model. J Agric Meteorol 73:2–8
Mathison C, Deva C, Falloon P et al (2017) Defining sowing and harvest dates based on the Asian summer monsoon. Earth Syst Dynam 9:563–592

Müller C, Elliott J, Chryssanthacopoulos J et al (2017) Global gridded crop model evaluation: benchmarking, skills, deficiencies and implications. Geosci Model Dev 10:1403–1422

Osborne TM, Wheeler TR (2013) Evidence for a climate signal in trends of global crop yield variability over the past 50 years. Environ Res Lett 8:024001. https://doi.org/10.1088/1748-9326/8/2/024001

Portmann FT, Siebert S, Döll P (2010) MIRCA2000—global monthly irrigated and rainfed crop areas around the year 2000: a new high-resolution data set for agricultural and hydrological modeling. Glob Biogeochem Cycles 24:GB1011. https://doi.org/10.1029/2008GB003435

Porwollik V, Müller C, Elliott J et al (2017) Spatial and temporal uncertainty of crop yield aggregations. Eur J Agron 88:10–21

Ray DK, Ramankutty N, Mueller ND et al (2012) Recent patterns of crop yield growth and stagnation. Nat Commun 3:1293. https://doi.org/10.1038/ncomms2296

Ray DK, Gerber JS, MacDonald GK et al (2015) Climate variation explains a third of global crop yield variability. Nat Commun 6:5989. https://doi.org/10.1038/ncomms6989

Rippey BR (2015) The U.S. drought of 2012. Weather Clim Extreme 10:57–64

Rosenzweig C, Hillel D (eds) (2008) Climate variability and the global harvest. Impacts of El Niño and other oscillations on agro-ecosystems. Oxford University Press, Oxford

Sacks WJ, Deryng D, Foley JA et al (2010) Crop planting dates: an analysis of global patterns. Glob Ecol Biogeogr 19:607–620

Schauberger B, Gornott C, Wechsung F (2017) Global evaluation of a semiempirical model for yield anomalies and application to within-season yield forecasting. Glob Chang Biol 23:4750–4764

See L, Fritz S, You L et al (2015) Improved global cropland data as an essential ingredient for food security. Glob Food Sec 4:37–45

Sivakumar MVK, Hansen J (eds) (2007) Climate prediction and agriculture. Advances and challenges. Springer, Berlin/Heidelberg

Ubilava D (2017) The ENSO effect and asymmetries in wheat price dynamics. World Dev 96:490–502

van der Velde M, Nisini L (2019) Performance of the MARS-crop yield forecasting system for the European Union: assessing accuracy, in-season, and year-to-year improvements from 1993 to 2015. Agric Syst 168:203–212

Waha K, van Bussel LGJ, Müller C et al (2012) Climate-driven simulation of global crop sowing dates. Glob Ecol Biogeogr 21:247–259

Welton G (2011) The impact of Russia's 2010 Grain Export Ban. Oxfam GB, Oxford. http://www.oxfam.org/sites/www.oxfam.org/files/rr-impact-russias-grain-export-ban-280611-en.pdf. Accessed 26 Nov 2018

Whitcraft KA, Becker-Reshef I, Justice OC (2015) A framework for defining spatially explicit earth observation requirements for a global agricultural monitoring initiative (GEOGLAM). Remote Sens 7:2072–4292

Wilks DS, Godfrey CM (2002) Diagnostic verification of the IRI net assessment forecasts, 1997–2000. J Clim 15:1369–1377

Yuan C, Yamagata T (2015) Impacts of IOD, ENSO and ENSO Modoki on the Australian winter wheat yields in recent decades. Sci Rep 5:17252. https://doi.org/10.1038/srep17252

Zampieri M, Ceglar A, Dentener F et al (2017) Wheat yield loss attributable to heat waves, drought and water excess at the global, national and subnational scales. Environ Res Lett 12:064008. https://doi.org/10.1088/1748-9326/aa723b

Chapter 8
Agricultural Adaptation to Climate Change in China

Zhan Tian, Hanqing Xu, Honglin Zhong, Laixiang Sun, and Junguo Liu

Abstract The climate warming and its impacts on agricultural production and food security have raised great public concerns in China. To mitigate the negative impact of climate change on agriculture, maintain crop production levels, and even identify opportunities in agricultural intensification, researchers have put great efforts to evaluate changes in agro-climatic resources and develop adaptation measures across different cropping zones of China under climate change. This chapter reviews the available literature on climate change impacts and adaptations in agriculture in China. The insights derived from this review are expected to be useful for decision-makers of the national government in the fields covering agricultural adaptation and food security.

8.1 Introduction

Global annual average temperature has increased at a rate of 0.19 °C per decade for the last 50 years and has displayed a strong warming trend in the twenty-first century (Pachauri et al. 2014). A warmer climate would negatively impact crop production and food security (Howden et al. 2007; Tripathi et al. 2016). Warming directly affects crop growth and yield through stress caused by heat stress and water deficit,

Z. Tian · J. Liu (✉)
School of Environmental Science and Engineering, Southern University of Science and Technology, Shenzhen, China
e-mail: liujg@sustech.edu.cn

H. Xu
Key Laboratory of Geographic Information Science (Ministry of Education), East China Normal University, Shanghai, China

H. Zhong
Department of Geographical Sciences, University of Maryland, College Park, MD, USA

L. Sun
Department of Geographical Sciences, University of Maryland, College Park, MD, USA

International Institute for Applied Systems Analysis (IIASA), Laxenburg, Austria

T. Iizumi et al. (eds.), *Adaptation to Climate Change in Agriculture*,
https://doi.org/10.1007/978-981-13-9235-1_8

and indirectly influences pathogens and weeds, and hence farmers' field management (Celikkol Erbas and Guven Solakoglu 2017; Elad and Pertot 2014). Those potential impacts of climate change have raised great public concerns and led to studies that evaluate the agricultural impacts and possible adaptation measures in China (Iglesias and Garrote 2015; Kettle et al. 2014).

The warming trend is likely irreversible in the short term, and subsequently adaptation is unavoidable. To mitigate the risk of climate-induced production reduction, researchers have developed various adaptation technologies, such as high-yielding varieties and water-saving practices (Hertel and Lobell 2014; Lobell 2014; Shiferaw et al. 2014).

Many studies have been carried out to help agricultural adaptations in China to climate warming, with various management practices at specific locations (Zhao et al. 2017, 2018). Assessments based on process-based crop models show that crop yields would decrease under future climate conditions when current optimal management is assumed (Ali et al. 2017). In contrast, potential opportunities (e.g., removal of cold temperature limitations, prolonged growing season, improved water-use efficiency, etc.) under climate change have received less attention (Jiang et al. 2007). A cropping system refers to a sequence of crops grown in the same field that aims at maximizing annual total production (or profit) per unit land area (Blanco-Canqui and Lal 2008). Current cropping systems would enable adaption to a warmer climate (Deng et al. 2017), and research focusing on cropping system adaptation could help identify potential benefits of warming (Beveridge et al. 2018).

There is an increasing research attention to the adaptation at regional scale, compared to those at the farm field scale. Examples include agricultural intensification and expansion of multi-cropping areas in a cropping zone (Brown et al. 2011; Kurukulasuriya and Rosenthal 2014; Petrie and Bates 2017; Tian et al. 2015). Given the historical prominence of the multi-cropping practice in China, it is intuitive to imagine that the multi-cropping practice could benefit from warming and increase annual total production in a given area (Tian et al. 2014). If this is true, the multi-cropping practice is promising as a regional-scale adaptation measure. However, a comprehensive review on the agricultural adaptation to climate change still remains rare (Jones et al. 2016).

This review focuses on the climate change impacts on production of major crops and agricultural adaptation in China. Various adaptation practices have been developed. Practices appearing in this review include use of new technologies to improve water-use efficiency and water availability and avoid soil erosion, utilization of new cropping systems, shift of cropping schedules, switch of crop varieties, and introduction of governmental policies and institutional programs. These measures have been successfully applied to moderate the negative impacts of climate change over the major cropping zones in China. The insights derived from this review would help ensure food security in China under climate warming and could be applied to other regions of the world.

8.2 Changes in Agro-Climatic Resources and Climate Disasters

The climate warming has changed agro-climatic resources and led to more frequent climate disasters which could affect crop growth and production in China.

8.2.1 Changes in Agro-Climatic Resources

Global warming alters temperature and precipitation, and exerts significant impacts on crop growth and production through changes in agro-climatic resources (Fischer et al. 2005). There are rising concerns on how climate change has altered and will continue to alter agro-climatic conditions in China (Fan et al. 2017; Kuo et al. 2015). As of 2019, China has experienced a temperature increase of 1.2 °C since 1960. Northeast China has shown a stronger increase of 0.36 °C per decade with a winter precipitation increase with an average rate about 16% per decade (Piao et al. 2010). It has been acknowledged that warming for the past 50 years has led to a prolonged growing season and northward expansion of multi-cropping systems in China (Yang et al. 2010). The growing season is prolonged by 1.16 days per year in the specific agro-climate zone. The green up of vegetation in spring is advanced by 0.79 days per year, whereas crop dormancy in autumn is delayed by 0.37 days per year (Piao et al. 2006). The annual number of frost days showed a decreasing trend in Northern China (Fu et al. 2015; Hallmann et al. 2017).

Agro-climatic resources have significantly changed during the last 50 years in China (Yin et al. 2016). It is reported that this change has impacted the potential yield, timing of planting, geographical distribution of multi-cropping area, and crop varieties used by farmers (Altieri et al. 2015; Jones et al. 2017; Ye et al. 2015). For example, due to temperature increase and precipitation change in the North China Plain, water availability in the multi-cropping systems currently operating is unsustainable and results in overpumping of groundwater used for irrigation to meet the crop requirement (Wang et al. 2017). Southwest China may experience a significant decrease in humidity, and crop production would face a severe drought risk (Song et al. 2012). In regions where agro-climatic resources have been improved, such as Northern China, a warmer climate has led to a longer growing season than previously. A more intense multi-cropping system can be introduced to increase production in higher latitudes (Tian et al. 2014). A spatially explicit understanding of changes in agro-climatic resources in the historical past and projected future could support policymakers, farming communities, and other stakeholders to evaluate production risk, plan adaptation and mitigation, and improve field management.

8.2.2 Changes in Climate Disasters

Climate disasters, such as drought, flood, tropical cyclone, cold damage, and heatwave are the main reasons for reduction in production (Lesk et al. 2016). The increase in seasonal and interannual climate fluctuations under climate change would aggravate agro-meteorological disasters (Huang et al. 2017).

Drought accounts for up to 30% of total agricultural losses associated with agro-meteorological disasters (Wu et al. 2011, 2016). Climate warming would change precipitation patterns and affect water resources and irrigation demand (Gornall et al. 2010). During the past three decades, drought affected a Crop land area of 23.9 million ha per year, on a multiyear average basis (Liu et al. 2015). Moreover, recent studies indicate that extreme drought events could become more frequent and intense in Northern China (Trenberth et al. 2014). A warmer climate would increase the drought risk at key crop growth stages (Yu et al. 2018), as drought events during the reproductive growth stage result in severe yield losses (Xu et al. 2019). Climate change would modulate water cycle, water resources, frequency of flood and drought, and crop production (Zhang and Hu 2018). To avoid drought damage during the early growth stages of maize (early spring), farmers have delayed the sowing date of spring maize in Northern China. Flood can be a local disaster due to the orographic effect. Compared to drought, only a limited number of studies are available for the influences of flood on crop production (Chen et al. 2018a). In recent years, the frequency of flood, which affects agricultural production, farmers' lives, and even national economic development, has rapidly increased (Yang et al. 2016; Zhang et al. 2015, 2016). While the frequency and intensity of flood in Southwest China have increased (Ji et al. 2015), those in Northeast China have decreased (Chou et al. 2019; Zhang et al. 2016). The frequency, intensity, and duration of flood in most parts of China are projected to increase (Huo-Po et al. 2013). It is necessary to address future flood risk in agriculture across different cropping zones of China under climate change.

Heatwaves are consecutive days with high temperatures exceeding a threshold that could lead to damage to crop production (Schauberger et al. 2017). Furthermore, heatwaves would increase soil and crop potential evapotranspiration (Karmakar et al. 2016). The impacts of heatwaves on agricultural production could be substantial (Wahid et al. 2007). Heat stress at the spring maize grain-filling stage is the main obstacle to increasing yield in the North China Plain (Tao et al. 2016). Heatwaves often accompany droughts, reducing yield and quality of grains (Fahad et al. 2017; Sehgal et al. 2018). Due to heatwaves, winter wheat and spring wheat under irrigated conditions at the end of this century may suffer from yield loss of 7.1% and 17.5%, respectively (Yang et al. 2017). In some future climate scenarios, precipitation decreases would accompany heatwaves (Zaitchik et al. 2006).

Low-temperature events have decreased in Northern China. However, the frost risk to spring maize in Northeast China has become higher because of the northward shift of cropland area and adoption of early-mid maturity varieties with earlier sowing dates (Zhao et al. 2015). The evaporative cooling under irrigated conditions would also increase the risk of low-temperature damage during the drought episodes in winter and spring seasons (Zhu et al. 2012).

8.3 Climate Change Impacts on Agriculture in China

Agro-climatic resources determine potential yield and extent of the suitable area for cropping. Climate change and climate disasters pose great risks for agriculture, with reduced opportunities. More detailed analyses to identify the climate change impacts on agro-climatic resources, cropping systems, and crop production at a local scale are required.

8.3.1 Impacts on Potential Yields

Climate change affects the potential yields of crops in China positively and negatively. Warmer climate would accelerate the growth and decrease yields of rice, maize, and wheat (Deng et al. 2010). In the arid and semiarid regions of China, a warmer climate and less precipitation reduce water resources and increase drought risk (Sivakumar et al. 2005). Liu et al. (2013) simulated the yields at ten locations in Northeast China by changing precipitation from –10% to –30% of the current level (1981–2010), resulting in great reductions of over 60% in many cases except for the locations with sufficient precipitation. On the other hand, crop yield in Northeast China is projected to increase due to reduced cold damage (Liu et al. 2013).

Climate change impacts precipitation patterns during crop duration, which differently affect water availability and crop water demand by (Gornall et al. 2010). A warmer climate also increases irrigation demand to meet crop water requirements (Zhang et al. 2011) and increases drought risk (Abtew and Melesse 2013). A prolonged growing season in Northeast China under warming increases irrigation demand (Tao and Zhang 2010).

Although elevated CO_2 concentrations can increase photosynthetic rate and water-use efficiency, net impacts on crop production under climate change are under debate. For example, potential yields of spring maize and summer maize in the North China Plain may decrease under elevated CO_2 concentrations (Guo et al. 2010). Wheat production under both rain-fed and irrigated conditions may be increased if the CO_2 fertilization is considered (Lin et al. 2005). Rice production would decrease without the CO_2 fertilization, but increase if both the CO_2 fertilization and expansion of rice-producing area are considered (McGrath and Lobell 2013). Wei et al. (2009) indicated that rice production in three regions (Sichuan Basin, Yangtze River Basin, and Huang-Huai-Hai Plain) would decrease under future climate conditions without crop adaptation.

8.3.2 *Impacts on Cropping Systems*

Climate warming has changed the agro-climatic resources, and the adaptability of existing crop types and varieties. These drive agronomic adjustments (Liu et al. 2010). Understanding agronomic adjustments would help facilitate the development of adaptation strategies at a larger spatial scale.

Because of warming and precipitation increase, a northward shift of cropland area in China has emerged. The boundary of winter wheat cropping in China moved northward and westward (Yang et al. 2010; Tian et al. 2015). This is mainly due to the improved thermal resources and more days with less frost risk. A further northward shift of cropland area in Northern China is projected (Jin et al. 2002). In addition to wheat, a northward shift of the rice cropping area in Northeast China is also observed. Under projected climate conditions, some regions with single maize cropping in Northern China would become suitable for crops with a higher thermal requirement, such as single cropping rice (Li and Wang 2010; Yun et al. 2005).

On the other hand, more intensive multi-cropping systems are also being adapted under a warmer climate to increase annual total production per cropland area. Many studies show that the multiple cropping index would significantly increase, as the suitable cropping area in China shifts northward. The centroid of the rice cropping area in China has shifted over 370 km northeastward between 1949 and 2010 (Kuo et al. 2015). Such a cropland shift accompanies more intensive rice cropping than before, such as the replacement of double rice cropping with triple rice cropping in Southern China and that of single rice cropping with double rice cropping in Central China. A use of more intensive rice cropping systems would increase the annual total rice production in China (Bingchan 2012).

Many studies claim that crop production in China would decrease under the projected climate without crop adaptation. Potential yields of winter wheat and summer maize in the Huang-Huai-Hai Plain (112–122°E, 32–41°N) have declined in the past 50 years (Tian et al. 2006). The cropping system adaptation mentioned earlier could increase annual total production per cropland area in most areas of China (Zhao et al. 2010) (Holanda et al. 1998). New crop varieties that could utilize agro-climatic resources more efficiently are also important in adaptation (Zhang et al. 2011a). Such adaptation measures offer opportunities to achieve production gains in China.

8.4 Adaptation Measures

Agricultural adaptation is central to improving the resilience of cropping systems and moderating the adverse effects of climate change. Without CO_2 fertilization, effective adaptation, and genetic improvement, the national averages of the potential yields of wheat, rice, maize, and soybean would decrease by 6.0%, 3.2%, 7.4% and 3.1%, respectively (Zhao et al. 2017). Given the increase in food demand, crop adaptation measures in China are essential, as food security is always one of the main policy objectives in China. Improving the yields of cereal crops is essential to keep pace with population growth under climate change (Bruinsma 2017; Chen et al. 2018b).

8.4.1 Crop Varieties

Production losses caused by climate change can be reduced by adopting new varieties that are more tolerant to suboptimal conditions than earlier varieties. In Northern China, farmers have started using middle- or late-maturity varieties and delaying the sowing date to avoid frost damage during the early growth stages (Peng et al. 2008). New late-maturity maize varieties are widely adopted in the North China Plain to fully utilize the improved thermal condition in late summer, leading to a 4–6% increase in total production of the wheat–maize system despite the decrease in wheat yield (Wang et al. 2012). In the arid and semiarid areas with precipitation decrease, such as Northwest China and Northern China, drought-tolerant varieties have been developed and applied (Zhu et al. 2013). A warmer winter in the North China Plain requires farmers to replace traditional winter wheat varieties with new ones with low vernalization requirements (Li et al. 2013). For the aggravation of pests and diseases under climate change, breeding varieties more tolerant to pests and diseases is also important (Donatelli et al. 2017).

8.4.2 New Crop Rotation Patterns

Studies show that crop rotation patterns could increase yields in major cropping zones in China (Fan et al. 2011). The photosynthetic productivity potential of current wheat–rice rotation in the Yangtze–Huaihe region shows a declining trend (Fan et al. 2011). By introducing a new crop rotation pattern, the crop productivity per unit solar radiation and unit thermal resources in Eastern China increases. In Jiangsu Province, a wheat–rice rotation pattern has gradually been replaced with a rapeseed–rice or wheat–maize rotation pattern. In Anhui Province, the planting areas of wheat–rice and wheat–maize rotation patterns have kept increasing, while the rapeseed–rice rotation planting area has kept decreasing (Lin et al. 2017).

8.5 Conclusions

The impacts of climate change on agriculture are critical to food security in China. In this chapter, we reviewed the agricultural production impacts associated with climate change and climatic hazards. We also mentioned recent research progress in

China, which is based on regional climate and crop model simulations, and exploration of adaptation measures. A potential way of moving forward is to develop a more comprehensive evaluation system of agricultural impacts under climate change, which integrates the mechanism of crop growth and development, region-specific adaptation measures, and accurate climate and crop measurements. In the face of uncertainties in future production, it is desired to strengthen the confidence of climate impact estimates, characterizing differences across climate and crop models, and quantifying their combined effects by using a multi-model ensemble approach. An evaluation system of climate change impacts on multiple aspects of crop production (e.g., potential productivity, region-specific adaptation measures, and comparative advantages and disadvantages across crops) has been gradually formed in China.

Climate change has dramatically altered the spatial and temporal distributions of agro-climatic resources in China. Environmental sustainability becomes increasingly important, especially in arid and semiarid cropping zones in China because of an increase in potential evapotranspiration under a warmer climate. Weather and climate extremes have become more frequent and brought severe damage to agricultural production, but quantitative assessment is lacking. Agricultural systems are complex because of interactions among weather, soil, crop, and field management. Most studies focus on a single aspect of an agro-ecosystem. A more systematic view of agro-ecosystems is necessary to achieve multiple goals for food security and environmental sustainability.

Adaptation plays an essential role in mitigating the climate change impacts on production across different cropping zones in China. Crop calendar, variety, cropping system, agronomic technique, and field management need to be adjusted to maintain production under climate change. In regions where thermal conditions have improved due to warming, there is a potential for agricultural intensification. Given the importance of local characteristics in agriculture, a more systematic simulation model for agricultural production and adaptation covering different cropping zones across China is crucial.

Acknowledgments This work was supported by the National Natural Science Foundation of China (Grant Nos. 51761135024, 41671113, and 41601049).

References

Abtew W, Melesse A (2013) Climate change and evapotranspiration. In: Evaporation and evapotranspiration. Springer, pp 197–202

Ali S, Liu Y, Ishaq M, Shah T, Ilyas A, Din I (2017) Climate change and its impact on the yield of major food crops: evidence from Pakistan. Foods 6:39

Altieri MA, Nicholls CI, Henao A, Lana MA (2015) Agro ecology and the design of climate change-resilient farming systems. Agron Sustain Dev 35:869–890

Beveridge L, Whitfield S, Challinor A (2018) Crop modelling: towards locally relevant and climate-informed adaptation. Clim Chang 147:475–489

Bingchan Z (2012) Research progress of impact of climate change on crop system in China. Chin Agric Sci Bull 2:062
Blanco-Canqui H, Lal R (2008) Cropping systems. In: Principles of soil conservation and management. Springer, Dordrecht, pp 165–193
Brown I, Poggio L, Gimona A, Castellazzi M (2011) Climate change, drought risk and land capability for agriculture: implications for land use in Scotland. Reg Environ Chang 11:503–518
Bruinsma J (2017) World agriculture: towards 2015/2030: an FAO study. Routledge, London
Celikkol Erbas B, Guven Solakoglu E (2017) In the presence of climate change, the use of fertilizers and the effect of income on agricultural emissions. Sustainability 9:1989
Chen H, Liang Z, Liu Y, Jiang Q, Xie S (2018a) Effects of drought and flood on crop production in China across 1949–2015: spatial heterogeneity analysis with Bayesian hierarchical modeling. Nat Hazards 92:525–541
Chen Q, Liu Y, Ge Q, Pan T (2018b) Impacts of historic climate variability and land use change on winter wheat climatic productivity in the North China Plain during 1980–2010. Land Use Policy 76:1–9
Chou J, Xian T, Dong W, Xu Y (2019) Regional temporal and spatial trends in drought and flood disasters in China and assessment of economic losses in recent years. Sustainability 11:55
Deng Z, Wang Q, Zhang Q, Qing J, Yang Q, Yuan Z, Liu W, Xu J (2010) Impact of climate warming and drying on food crops in northern China and the countermeasures. Acta Ecol Sin 30:6278–6288
Deng A, Chen C, Feng J, Chen J, Zhang W (2017) Cropping system innovation for coping with climatic warming in China. Crop J 5:136–150
Donatelli M, Magarey RD, Bregaglio S, Willocquet L, Whish JPM, Savary S (2017) Modelling the impacts of pests and diseases on agricultural systems. Agric Syst 155:213–224
Elad Y, Pertot I (2014) Climate change impacts on plant pathogens and plant diseases. J Crop Improv 28:99–139
Fahad S, Bajwa AA, Nazir U, Anjum SA, Farooq A, Zohaib A, Sadia S, Nasim W, Adkins S, Saud S (2017) Crop production under drought and heat stress: plant responses and management options. Front Plant Sci 8:1147
Fan M, Shen J, Yuan L, Jiang R, Chen X, Davies WJ, Zhang F (2011) Improving crop productivity and resource use efficiency to ensure food security and environmental quality in China. J Exp Bot 63:13–24
Fan D, Ding Q, Tian Z, Sun L, Fischer G (2017) A cross-scale model coupling approach to simulate the risk-reduction effect of natural adaptation on soybean production under climate change. Hum Ecol Risk Assess 23:426–440
Fischer G, Shah M, Tubiello FN, Van Velhuizen H (2005) Socio-economic and climate change impacts on agriculture: an integrated assessment, 1990–2080. Philos Trans R Soc Lond B Biol Sci 360:2067–2083
Fu YH, Zhao H, Piao S, Peaucelle M, Peng S, Zhou G, Ciais P, Huang M, Menzel A, Peñuelas J (2015) Declining global warming effects on the phenology of spring leaf unfolding. Nature 526:104
Gornall J, Betts R, Burke E, Clark R, Camp J, Willett K, Wiltshire A (2010) Implications of climate change for agricultural productivity in the early twenty-first century. Philos Trans R Soc B Biol Sci 365:2973–2989
Guo R, Lin Z, Mo X, Yang C (2010) Responses of crop yield and water use efficiency to climate change in the North China Plain. Agric Water Manag 97:1185–1194
Hallmann CA, Sorg M, Jongejans E, Siepel H, Hofland N, Schwan H, Stenmans W, Müller A, Sumser H, Hörren T (2017) More than 75 percent decline over 27 years in total flying insect biomass in protected areas. PLoS One 12:e0185809
Hertel TW, Lobell DB (2014) Agricultural adaptation to climate change in rich and poor countries: current modeling practice and potential for empirical contributions. Energy Econ 46:562–575

Holanda F, Mengel D, Paula M, Carvaho J, Bertoni J (1998) Influence of crop rotations and tillage systems on phosphorus and potassium stratification and root distribution in the soil profile. Common Soil Sci Plant Anal 29:2383–2394

Howden SM, Soussana J-F, Tubiello FN, Chhetri N, Dunlop M, Meinke H (2007) Adapting agriculture to climate change. Proc Natl Acad Sci 104:19691–19696

Huang J, Lei Y, Zhang F, Hu Z (2017) Spatio-temporal analysis of meteorological disasters affecting rice, using multi-indices, in Jiangsu province, Southeast China. Food Sec 9:661–672

Huo-Po C, Jian-Qi S, Xiao-Li C (2013) Future changes of drought and flood events in China under a global warming scenario. Atmos Oceanic Sci Lett 6:8–13

Iglesias A, Garrote L (2015) Adaptation strategies for agricultural water management under climate change in Europe. Agric Water Manag 155:113–124

Ji Y, Zhou G, Wang S, Wang L (2015) Increase in flood and drought disasters during 1500–2000 in Southwest China. Nat Hazards 77:1853–1861

Jiang J, Xu X-B, Yayashi Y, F-l T (2007) Overview of the impacts of climate warming on paddy rice production in Heilongjiang province, the north most of China. Tsukuba Geoenviron Sci 3:23–31

Jin Z-Q, Ge D-K, Shi C-L (2002) Several strategies of food crop production in the Northeast China plain for adaptation to global climate change-a modeling study. Acta Agron Sin 28:24–31

Jones KR, Watson JE, Possingham HP, Klein CJ (2016) Incorporating climate change into spatial conservation prioritization: a review. Biol Conserv 194:121–130

Jones JW, Antle JM, Basso B, Boote KJ, Conant RT, Foster I, Godfray HCJ, Herrero M, Howitt RE, Janssen S (2017) Toward a new generation of agricultural system data, models, and knowledge products: state of agricultural systems science. Agric Syst 155:269–288

Karmakar R, Das I, Dutta D, Rakshit A (2016) Potential effects of climate change on soil properties: a review. Sci Int 4:51

Kettle NP, Dow K, Tuler S, Webler T, Whitehead J, Miller KM (2014) Integrating scientific and local knowledge to inform risk-based management approaches for climate adaptation. Clim Risk Manag 4:17–31

Kuo L, Dawei Z, Yanan H, Jianyong M, Yinlong X, Wilkes A, Opitz-Stapleton S, Lair K, Nadin R (2015) Feeding China. In: Climate risk and resilience in China. Routledge, London, pp 149–175

Kurukulasuriya P, Rosenthal S (2014) Climate change and agriculture: a review of impacts and adaptations. J Integr Agric 10:657–659

Lesk C, Rowhani P, Ramankutty N (2016) Influence of extreme weather disasters on global crop production. Nature 529:84

Li Y, Wang C (2010) Impacts of climate change on crop planting structure in China. Adv Clim Chang Res 6:123–129

Li K, Yang X, Mu C, Xu H, Chen F (2013) The possible effects of global warming on cropping systems in China VII-the effects of climate change on planting boundaries of different winter-spring varieties of winter wheat in China. Sci Agric Sin 46:1583–1594

Lin E, Xiong W, Ju H, Xu Y, Li Y, Bai L, Xie L (2005) Climate change impacts on crop yield and quality with CO2 fertilization in China. Philos Trans R Soc Lond Ser B Biol Sci 360:2149–2154

Lin K, Li D, Song X, Zu C, Jiang C, Shen J, Ma C, Gao Q, Zhu Q, Ji X (2017) Contents and changes of potassium in plough layers of Xuancheng, South Anhui Province. Agric Sci 8:348

Liu Y, Wang E, Yang X, Wang J (2010) Contributions of climatic and crop varietal changes to crop production in the North China Plain, since 1980s. Glob Chang Biol 16:2287–2299

Liu Z, Hubbard KG, Lin X, Yang X (2013) Negative effects of climate warming on maize yield are reversed by the changing of sowing date and cultivar selection in Northeast China. Glob Chang Biol 19:3481–3492

Liu Y, Pan Z, Zhuang Q, Miralles DG, Teuling AJ, Zhang T, An P, Dong Z, Zhang J, He D (2015) Agriculture intensifies soil moisture decline in northern China. Sci Rep 5:11261

Lobell DB (2014) Climate change adaptation in crop production: beware of illusions. Glob Food Sec 3:72–76

McGrath JM, Lobell DB (2013) Regional disparities in the CO2 fertilization effect and implications for crop yields. Environ Res Lett 8:014054

Pachauri RK, Allen MR, Barros VR, Broome J, Cramer W, Christ R, Church JA, Clarke L, Dahe Q, Dasgupta P (2014) Climate change 2014: synthesis report. Contribution of working groups I, II and III to the fifth assessment report of the Intergovernmental Panel on Climate Change. IPCC
Peng S, Tang Q, Zou Y (2008) Current status and challenges of Rice production in China. Plant Production Science 12:3–8
Petrie CA, Bates J (2017) 'Multi-cropping', intercropping and adaptation to variable environments in Indus South Asia. J World Prehist 30:81–130
Piao S, Fang J, Zhou L, Ciais P, Zhu B (2006) Variations in satellite-derived phenology in China's temperate vegetation. Glob Chang Biol 12:672–685
Piao S, Ciais P, Huang Y, Shen Z, Peng S, Li J, Zhou L, Liu H, Ma Y, Ding Y (2010) The impacts of climate change on water resources and agriculture in China. Nature 467:43
Schauberger B, Archontoulis S, Arneth A, Balkovic J, Ciais P, Deryng D, Elliott J, Folberth C, Khabarov N, Müller C (2017) Consistent negative response of US crops to high temperatures in observations and crop models. Nat Commun 8:13931
Sehgal A, Sita K, Siddique KH, Kumar R, Bhogireddy S, Varshney RK, Hanumantha Rao B, Nair RM, Prasad PV, Nayyar H (2018) Drought or/and heat-stress effects on seed filling in food crops: impacts on functional biochemistry, seed yields, and nutritional quality. Front Plant Sci 9:1705
Shiferaw B, Tesfaye K, Kassie M, Abate T, Prasanna B, Menkir A (2014) Managing vulnerability to drought and enhancing livelihood resilience in sub-Saharan Africa: technological, institutional and policy options. Weather Climate Extremes 3:67–79
Sivakumar M, Das H, Brunini O (2005) Impacts of present and future climate variability and change on agriculture and forestry in the arid and semi-arid tropics. In: Increasing climate variability and change. Springer, pp 31–72
Song Y, Liu Y, Ding Y (2012) A study of surface humidity changes in China during the recent 50 years. Acta Meteor Sin 26:541–553
Tao F, Zhang Z (2010) Adaptation of maize production to climate change in North China Plain: quantify the relative contributions of adaptation options. Eur J Agron 33:103–116
Tao Z, Chen Y, Chao L, Zou J, Peng Y, Yuan S, Xia W, Peng S (2016) The causes and impacts for heat stress in spring maize during grain filling in the North China Plain—a review. J Integr Agric 15:2677–2687
Tian Z, Liu J-y, Cao M-k (2006) Simulation of the impact of climate change on Chinese wheat production in Huang-Huai-Hai plain. J Nat Resour 21:598–607
Tian Z, Yang X, Sun L, Fischer G, Liang Z, Pan J (2014) Agro climatic conditions in China under climate change scenarios projected from regional climate models. Int J Climatol 34:2988–3000
Tian Z, Liang Z, Sun L, Zhong H, Qiu H, Fischer G, Zhao S (2015) Agriculture under climate change in China: mitigate the risks by grasping the emerging opportunities. Hum Ecol Risk Assess 21:1259–1276
Trenberth KE, Dai A, Van Der Schrier G, Jones PD, Barichivich J, Briffa KR, Sheffield J (2014) Global warming and changes in drought. Nat Clim Chang 4:17
Tripathi A, Tripathi DK, Chauhan D, Kumar N, Singh G (2016) Paradigms of climate change impacts on some major food sources of the world: a review on current knowledge and future prospects. Agric Ecosyst Environ 216:356–373
Wahid A, Gelani S, Ashraf M, Foolad MR (2007) Heat tolerance in plants: an overview. Environ Exp Bot 61:199–223
Wang J, Wang E, Yang X, Zhang F, Yin H (2012) Increased yield potential of wheat-maize cropping system in the North China Plain by climate change adaptation. Clim Chang 113:825–840
Wang Z, Xie P, Lai C, Chen X, Wu X, Zeng Z, Li J (2017) Spatiotemporal variability of reference evapotranspiration and contributing climatic factors in China during 1961–2013. J Hydrol 544:97–108
Wu J, He B, Lü A, Zhou L, Liu M, Zhao L (2011) Quantitative assessment and spatial characteristics analysis of agricultural drought vulnerability in China. Nat Hazards 56:785–801

Wu X, Lu Y, Zhou S, Chen L, Xu B (2016) Impact of climate change on human infectious diseases: empirical evidence and human adaptation. Environ Int 86:14–23

Wei X, Declan C, Erda L, Yinlong X, Hui J, Jinhe J, Yan L (2009) Future cereal production in China: the interaction of climate change, water availability and socio-economic scenarios. Global Environmental Change, 19(1), 34–44.

Xu H, Tian Z, He X, Wang J, Sun L, Fischer G, Fan D, Zhong H, Wu W, Pope E (2019) Future increases in irrigation water requirement challenge the water-food nexus in the northeast farming region of China. Agric Water Manag 213:594–604

Yang X, Liu Z, Chen F (2010) The possible effects of global warming on cropping systems in China I. The possible effects of climate warming on northern limits of cropping systems and crop yields in China. Sci Agric Sin 43:329–336

Yang J, Huo Z, Wu L, Wang T, Zhang G (2016) Indicator-based evaluation of spatiotemporal characteristics of rice flood in Southwest China. Agric Ecosyst Environ 230:221–230

Yang X, Tian Z, Sun L, Chen B, Tubiello FN, Xu Y (2017) The impacts of increased heat stress events on wheat yield under climate change in China. Clim Chang 140:605–620

Ye Q, Yang X, Dai S, Chen G, Li Y, Zhang C (2015) Effects of climate change on suitable rice cropping areas, cropping systems and crop water requirements in southern China. Agric Water Manag 159:35–44

Yin X, Olesen JE, Wang M, Öztürk I, Zhang H, Chen F (2016) Impacts and adaptation of the cropping systems to climate change in the northeast farming region of China. Eur J Agron 78:60–72

Yu H, Zhang Q, Sun P, Song C (2018) Impact of droughts on winter wheat yield in different growth stages during 2001–2016 in Eastern China. Int J Disaster Risk Sci 9:376–391

Yun Y, Fang X, Wang Y, Tao J, Qiao D (2005) Main grain crops structural change and its climate background in Heilongjiang province during the past two decades. J Nat Resour 20:697–704

Zaitchik BF, Macalady AK, Bonneau LR, Smith RB (2006) Europe's 2003 heat wave: a satellite view of impacts and land–atmosphere feedbacks. Int J Climatol 26:743–769

Zhang Q, Hu Z (2018) Assessment of drought during corn growing season in Northeast China. Theor Appl Climatol:1–7

Zhang Q, Xu C-Y, Chen X (2011) Reference evapotranspiration changes in China: natural processes or human influences? Theor Appl Climatol 103:479–488

Zhang F, Cui Z, Fan M, Zhang W, Chen X, Jiang R (2011a) Integrated soil–crop system management: reducing environmental risk while increasing crop productivity and improving nutrient use efficiency in China. J Environ Qual 40:1051–1057

Zhang Q, Gu X, Singh VP, Kong D, Chen X (2015) Spatiotemporal behavior of floods and droughts and their impacts on agriculture in China. Glob Planet Chang 131:63–72

Zhang Q, Gu X, Singh VP, Liu L, Kong D (2016) Flood-induced agricultural loss across China and impacts from climate indices. Glob Planet Chang 139:31–43

Zhao J, Yang X, Liu Z, Cheng D, Wang W, Chen F (2010) The possible effect of global climate changes on cropping systems boundary in China II. The characteristics of climatic variables and the possible effect on northern limits of cropping systems in South China. Sci Agric Sin 43:1860–1867

Zhao J, Yang X, Dai S, Lv S, Wang J (2015) Increased utilization of lengthening growing season and warming temperatures by adjusting sowing dates and cultivar selection for spring maize in Northeast China. Eur J Agron 67:12–19

Zhao C, Liu B, Piao S, Wang X, Lobell DB, Huang Y, Huang M, Yao Y, Bassu S, Ciais P (2017) Temperature increase reduces global yields of major crops in four independent estimates. Proc Natl Acad Sci 114:9326–9331

Zhao J, Zhan X, Jiang Y, Xu J (2018) Variations in climatic suitability and planting regionalization for potato in northern China under climate change. PLoS One 13:e0203538

Zhu X, Liang S, Pan Y (2012) Observational evidence of the cooling effect of agricultural irrigation in Jilin, China. Clim Chang 114:799–811

Zhu X, Li Y, Li M, Pan Y, Shi P (2013) Agricultural irrigation in China. J Soil Water Conserv 68:147A–154A

Part III
Adaptation Practices

Chapter 9
Agricultural Adaptation Policy in Japan

Shinya Yuji

Abstract This chapter presents an English translation of the section on agricultural impacts, which is a part of the Assessment Report on the Impact of Climate Change in Japan originally published in Japanese, 2015. The translated document enables researchers who are not accessible to the original report written in Japanese capturing an overview of the observed agricultural impacts in Japan due to climate change and associated governmental policy responses to adapt to climate change. Such information helps lead to a standardized international comparison in agricultural adaptation policy. The subsection numbers in this chapter follow those used in the Climate Change Impact Assessment Report.

9.1 Summary of Japan's Climate Change Impact Assessments

Japan has set up a subcommittee on climate change impact assessment under the Global Environmental Committee of the Central Environment Council in July 2013 and has been conducting deliberations in order to assess the impact of climate change in Japan. The subcommittee compiled the results of deliberations as "Assessment Report on the Impact of Climate Change in Japan" in March 2015. Also, the Central Environment Council expressed its opinion to the Minister of the Environment as "Report on impact assessment of climate change in Japan and future issues (hereafter referred to as the Climate Change Impact Assessment Report)."

This section presents summaries of Japan's climate change impact assessments for each of the seven sectors indicated in the Climate Change Impact Assessment Report (agriculture, forest/forestry and fishery; water environment/water resources; natural ecosystems; natural disasters/coastal areas; human health; industrial and

S. Yuji (✉)
Environment Policy Office, Minister's Secretariat, Ministry of Agriculture, Forestry and Fisheries of Japan, Tokyo, Japan
e-mail: shinya_yuji370@maff.go.jp

T. Iizumi et al. (eds.), *Adaptation to Climate Change in Agriculture*,
https://doi.org/10.1007/978-981-13-9235-1_9

economic activities; and citizenry life/urban life).

For the impact assessments, expert judgment based on scientific findings was used to rate the three parameters of "significance," "urgency," and "certainty" from the following perspectives Table 9.1.

The impact assessment results are expressed using the following legend.

Legend				
(Significance)	● Very high	◆ Intermediate		– N/A (cannot be assessed at present)
(Urgency)	● High	▲ Medium	□ Low	– N/A (cannot be assessed at present)
(Certainty)	● High	▲ Medium	□ Low	– N/A (cannot be assessed at present)

Table 9.1 Summary of the findings of Japan's climate change impact

Sectors	Categories	Subcategories	Significance	Urgency	Certainty
Agriculture, forest/forestry, fishery	Agriculture	Paddy field rice	●	●	●
		Vegetables	–	▲	▲
		Fruit trees	●	●	●
		Barley/wheat, soybean, feed crops, and other crops	●	▲	▲
		Livestock farming	●	▲	▲
		Plant pests and weeds	●	●	●
		Water, land, and agricultural infrastructure	●	●	▲
	Forest/forestry	Timber production (e.g., plantations)	●	●	□
		Non-wood forest products (e.g., mushrooms)	●	●	□
	Fishery	Migratory fish stocks (ecology of fishes)	●	●	▲
		Propagation and aquaculture	●	●	□
Water environment, water resources	Water environment	Lakes/marshes, dams (reservoir)	●	▲	▲
		Rivers	◆	□	□
		Coastal areas and closed sea areas	◆	▲	□
	Water resources	Water supply (surface water)	●	●	▲
		Water supply (groundwater)	◆	▲	□
		Water demand	◆	▲	▲

(continued)

Table 9.1 (continued)

Sectors	Categories	Subcategories	Significance	Urgency	Certainty
Natural ecosystems	Terrestrial ecosystems	Alpine/subalpine zone	●	●	▲
		Natural forests/ secondary forests	●	▲	●
		Countryside-landscape (Satochi-Satoyama)	◆	▲	□
		Planted forests	●	▲	▲
		Damage from wildlife	●	●	–
		Material balance	●	▲	▲
	Freshwater ecosystems	Lakes/marshes	●	▲	□
		Rivers	●	▲	□
		Marshlands	●	▲	□
	Coastal ecosystems	Subtropics	●	●	▲
		Temperate/subarctic	●	●	▲
	Marine ecosystems		●	▲	□
	Phenology		◆	●	●
	Shifts in distribution and populations		●	●	●
Natural disasters, coastal areas	Rivers	Floods	●	●	●
		Inland waters	●	●	▲
	Coastal areas	Sea-level rise	●	▲	●
		Storm surges, high waves	●	●	●
		Coastal erosion	●	▲	▲
	Mountain areas	Debris flows, landslide, and other disasters	●	●	▲
	Others	Strong winds and other	●	▲	▲
Human health	Winter warming	Mortality in winter season	◆	□	□
	Heat stress	Risk of mortality	●	●	●
		Heat illness	●	●	●
	Infection	Water- and food-borne diseases	–	–	□
		Vector-borne diseases	●	▲	▲
		Other infectious diseases	–	–	–
	Others		–	▲	▲

(continued)

Table 9.1 (continued)

Sectors	Categories	Subcategories	Significance	Urgency	Certainty
Industrial/ economic activities	Manufacture		◆	□	□
	Energy	Energy demand and supply	◆	□	▲
	Commerce		–	–	□
	Finance, insurance		●	▲	□
	Tourism	Leisure	●	▲	●
	Construction		–	–	–
	Medical		–	–	–
	Others	Other impacts (e.g., overseas impact)	–	–	□
Citizenry life, urban life	Urban infrastructure, critical services	Water supply, transportation, and other	●	●	□
	Life with sense of culture and history	Phenology, traditional events/ local industry	◆	●	●
	Others	Impacts on life due to heat stress	–	●	□
	Urban infrastructure, critical services	Water supply, transportation, and other	●	●	●

9.2 Basic Adaptation Measures

9.2.1 Formulation of Adaptation Plan

The Climate Change Impact Assessment Report formulated in March 2015 showed that the impacts of climate change are beginning to appear in Japan, including an increase in air temperature and frequency of heavy rainfall, a decrease in number of days with precipitation, and an increase in sea surface temperature. It also showed that due to high temperature, impacts can already be seen, including a decline in quality of agricultural crops and shifts in the distribution of flora and fauna. It also revealed that in the future, there may be a further increase in temperature and the frequency of heavy rainfall events, a decrease in the number of days of precipitation, and an increase in sea surface temperature, as well as an increase in the amount of precipitation per heavy rainfall events, an increase in the maximum intensity of typhoons, and sea-level rise. The report reveals that a variety of impacts may occur in various sectors, including agriculture, forestry, fishery, the water environment, water resources, natural ecosystems, natural disasters, and human health.

In order to promote systematic and integrated efforts to address these impacts of climate change, the first Adaptation Plan was approved by the Cabinet in November 2015. The plan established basic principles, such as a future vision of society, as well as basic approaches, basic principles, basic measures, and international measures for each sector.

The Japanese Diet promulgated the Climate Change Adaptation Act (Act No. 50 of 2018) in June 2018. This new law aims at legally defining adaptation strategies and promoting them in cooperation with local governments and private sector. National Adaptation Plan required by this law was approved by the Cabinet in November 2018. Section 9.2.2 is based on the National Adaptation Plan (2018).

9.2.2 *Basic Adaptation Measures for Agriculture*

9.2.2.1 Overview of Agricultural Production

Various efforts have been made in the field of agricultural production in general. For instance, adaptation measures are being disseminated to agricultural producers. The measures include adoption of technology aimed at avoiding or mitigating impacts of high temperature, etc. and introduction of plant varieties that are tolerant to high temperatures. Also, demonstration of new adaptation technologies is underway.

Furthermore, in cooperation with local governments (or related institutions), impacts of global warming are monitored and information related to adaptation measures are offered in the "Global Warming Impact Investigation Report" and on the website of the Ministry of Agriculture, Forestry and Fisheries.

Special efforts will be made for paddy rice, fruit trees, plant pests, and weeds whose significance, urgency, and certainty were estimated to be high in the Climate Change Impact Assessment Report.

For other categories, in addition to the continuance of previously taken measures, development of new adaptive varieties and cultivation management technologies as well as basic research required for the development will be promoted regarding perspectives of future impacts.

Moreover, in cooperation with local governments (or related institutions), impacts of global warming will be monitored, and information related to adaptation measures will continue to be offered in the "Global Warming Impact Investigation Report" and on the website of the Ministry of Agriculture, Forestry and Fisheries.

9.2.2.2 Paddy Rice

Active efforts are being made to establish basic technologies of soil and water management to adapt to high temperature and to disseminate high-temperature-tolerant varieties. Although the cultivation area of high-temperature-tolerant varieties has increased gradually, such varieties have not been widely adopted due to the disagreement with actual consumer needs (according to the Global Warming Impact Investigation Report 2017, the planted area percentage of high-temperature-tolerant varieties was approximately 6.8%[1]).

[1] Rice-paddy acreage all over the country in 2017 was 1,370,000 ha, and the planted area of high-temperature-tolerant varieties was 93,800 ha.

For pest control, it has been promoted to take pest prevention measures at appropriate timings by utilizing forecast information.

In addition to the efforts shown above, the following measures will be taken hereafter:

For the development of new varieties, varieties with high-temperature tolerance that are less likely to be damaged by high temperature should basically be selected.

A decrease in yield has already been seen in years with extremely high temperature. Since much higher temperature is expected in the future, developing breeding materials capable of maintaining fertility under high-temperature sterility[2] will be promoted in order to prevent a decrease in yield.

Efforts will be continued to establish basic technologies of soil and water management. Support is to be provided to efforts such as demonstration to spread high-temperature-tolerant varieties, with all the stakeholders such as producers and actual consumers working together.

Promotion of adjusted pest control utilizing forecasting information is to be continued. By around 2019, damage mitigation technologies against pests will be developed and disseminated. The pests include rice sheath blight disease and rice streak, which are expected to increase due to global warming.

9.2.2.3 Fruit Trees

In order to mitigate sunburn damage on satsuma mandarin orange fruits caused due to high temperature and strong solar radiation, thinning out the fruits on of the upper part of the tree crown which is exposed to direct sunlight has been promoted. To mitigate peel puffing damage on the fruit, application of plant growth regulators such as calcium compounds has been promoted. Moreover, for measure against poor coloring, spraying of Figaron (ethychlozate),[3] which is commonly used for fruit thinning, has been promoted.

It is also promoted to switch from satsuma mandarin oranges to medium late ripening citrus fruits (such as *Citrus reticulata* Shiranui and blood orange) that prefer warm climate.

Regarding apples, in addition to introducing superior colored varieties including “Akibae” and yellow-colored varieties as measures against poor coloring, introduction of irrigation and reflection sheets have been promoted as measures against sunburn and poor coloring of the fruit.

As common measures against drought for fruits including peaches and cherries, the following have been promoted: water evaporation control with mulch, deep plowing and feeding of organic substrates at the dormant stage in order to maintain

[2] Due to high temperature during the flowering period, pollination is hindered and starch does not accumulate in the grain.

[3] Plant growth regulator used for the purpose of accelerating the ripening period, fruit thinning, and mitigating peel puffing of citrus fruits

soil moisture, and taking timely prevention measures for spider mites and other pests that are likely to occur in drought. Also, as a measure against frost damage due to late frost during the flowering period, establishment of a frost damage alert system has been promoted.

As a common measure against poor coloring of fruit due to climate change, a production and distribution system for fruits for processing use is being established in order to positively utilize such fruits as ingredients for fruit juice.

In addition to the efforts shown above, the following measures will be taken hereafter:

Regarding satsuma mandarin oranges, dissemination of the following cultivation management technologies should be accelerated: spraying of gibberellin[4] combined with prohydrojasmon[5] that can mitigate peel puffing, utilizing light-shielding materials that prevent sunburn of the fruit. For the purpose of stabilizing the flower setting, development of production-stabilizing technologies is to be commenced by improving fertilizer application and water management.

Regarding apples, development of cultivation management technologies will be promoted in order to reduce poor coloring and sunburn at high temperature.

Regions suitable for apple production are likely to shift in the future. Support will be provided for cultivation demonstration in regions with higher elevation and for switching varieties.

In the case of grapes, the following measures will be taken. Against poor coloring, introduction of superior colored varieties such as "Grosse Krone" and yellow-green-colored varieties such as "Shine Muscat" will be promoted. To mitigate coloring trouble due to high temperature at the ripening stage, dissemination of production-stabilizing technologies such as girdling[6] will be promoted.

In the case of Japanese pears, the following measures will be taken to mitigate damage caused by poor sprout emergence: utilization of sprout promoters, further introduction and dissemination of technical measures such as changing the application timing of fertilizers, and the implementation of developing production-stabilizing technologies in warm regions through measures including soil improvement.

In terms of breeding of satsuma mandarin oranges, apples, and Japanese pears, breeding materials adaptive to high-temperature conditions will be developed by or around 2019. The new varieties will subsequently be cultivated and introduced in production sites for demonstration.

In addition, if global warming advances due to climate change, it is expected that the area where protected cultivation of subtropical and tropical fruits is possible will

[4] Plant growth regulator used for the purposes including accelerating the growth of fruit trees, accelerating flowering and fruit enlargement

[5] Plant growth regulator used for the purposes including accelerating coloring of fruit and mitigating peel puffing of satsuma mandarin oranges

[6] A technique leading to improvement of coloring as follows: peeling off the outer skin of a trunk and sending nutrients made in leaves to a fruit cluster, without sending them to parts lower than the peeled part

expand. Therefore, subtropical and tropical fruits with high added value (such as atemoyas, avocados, mangoes, and lychees) will be introduced for demonstration, and regarding the producers' preference, shifts from the existing fruits will be promoted.

If the temperature zone suitable for apple cultivation moves northward due to the advancement of global warming, it is considered possible to form production areas in a new region. When forming such new production sites, establishment of low-cost and labor-saving orchards will be promoted.

Fruit trees are perennial crops, so they require a certain period of time to bear fruit, and the prices of fruits tend to fluctuate due to supply and demand imbalances. Therefore, it is especially necessary to implement measures from a long-term perspective compared to other crops. It is essential to establish networks among major production regions and prefectures, in order to share precise information such as global warming impacts and adaptive measures to examine action plans at production sites Fig. 9.1.

Fig. 9.1 High-temperature injury of agricultural production. (Upper left) Cross-section of paddy field rice ((left) white immature grain and (right) normal grain); (upper right) mandarin citrus peel puffing; (lower left) poor coloring of apple; and (lower right) poor coloring of grape

9.2.2.4 Land-Extensive Crops

As for wheat and barley, the following measures have been taken against heavy rainfall and moisture excess damage: dissemination of basic drainage techniques, prevention of *Fusarium* head blight at an appropriate timing, harvesting at an appropriate timing, and shifting to varieties resistant against diseases including *Fusarium* head blight and against preharvest sprouting.[7] As a result, certain effects have been observed. Moreover, as measures against frost damage, efforts have been taken to develop and disseminate varieties and breeding materials adaptive to climate change, as well as production-stabilizing technologies.

For soybeans, the following measures have been taken against heavy rainfall, high temperature, and drought: widely promoting drainage measures and the disseminating a groundwater level control system. As a result, certain effects have been observed. In addition, as pest control, efforts have been taken to develop and disseminate varieties and breeding materials resistant to pests as well as weed control techniques. Further efforts are being made to develop a cultivating system less likely to be impacted by climate change, such as the application of organic manure and crop rotation, which can mitigate the risk of pest and disease.

In the case of red beans, the dissemination of a high-temperature-tolerant variety, "Kitaasuka," has been promoted in Hokkaido (central Hokkaido and southern Hokkaido).

Regarding tea plants, a measure against frost damage including the introduction of a frost protection technique utilizing an energy-saving frost-protective fan system has been promoted. As a result, certain effects have been observed. Moreover, soil moisture evaporation control by grass mulch and sprinkling water has been implemented against drought. As for pest control, utilization of disease projection information, and replacing existing trees with varieties resistant to white peach scale[8] has been promoted.

Regarding sugar beet, development and dissemination of species resistant to diseases caused by high temperature have been carried out. As a result, certain effects have been observed. For high temperature, periodical monitoring and investigation of the production status are to be conducted, and findings will be accumulated in order to select the most suitable species. Moreover, drainage measures against heavy rain are being taken.

In the case of potatoes, removing or stomping snow after harvest have been promoted in Hokkaido to prevent leftover potatoes from sprouting the next spring and becoming weeds. Freezing of the soil and tuber frost-killing have been promoted as well.

Measures taken so far will be continuously promoted hereafter.

---

[7] A phenomenon where a sprout comes out of a seed that grew on an ear due to events including rainfall before harvesting

[8] A major pest insect for tea, which is a parasite living in the branches and trunk of a tea plant. It causes damages including dieback, due to decaying tree vigor. In recent years, its outbreak has been frequent nationwide, but there is no clear cause-and-effect relationship between climate change and the trend.

9.2.2.5 Horticultural Crops

Regarding vegetables, as measures against high-temperature conditions, development and dissemination of breeding materials adaptive to warmer conditions are being promoted. Efforts have been made to stabilize supply of open-field vegetables by selecting adequate varieties, adjusting the cultivation periods, and conducting pest control at appropriate timings. The following drought measures will be promoted: improvement of irrigation facilities, soil moisture evaporation control with mulch, and a timely extermination of spider mites and other pests that are likely to occur in drought.

Regarding ornamental plants, as measures against high-temperature conditions, adequate use of sprinkling water and the dissemination of varieties adaptive to hot conditions have been promoted.

Regarding vegetables and ornamental plants grown in greenhouses, as measures against high-temperature conditions, in addition to adequate air ventilation and shading, the introduction of mulch for soil temperature control, fog cooling system, pads and fans,[9] circulating fans, and heat pumps[10] and so on are being promoted. The following measures against typhoons and heavy snow are promoted: implementation of low-cost weather-resistant greenhouses, reinforcement of pipe-framed greenhouses, and installation of backup power sources. As a result, certain effects have been observed.

Measures taken so far will be continuously promoted hereafter.

9.2.2.6 Livestock Farming

As for livestock, measures have been taken to secure adequate barn environment by disseminating summer heat measures, such as sprinkling water and misting inside the barn, as well as applying lime and sprinkling water on the roof. Adequate feeding management techniques include avoiding close rearing and frequently shearing hair and supplying cold water and feedstuff of good quality. Moreover, efforts have been taken to develop and disseminate productivity improvement techniques to prevent a decline in the rate of gain and fertility in the summertime through adequate nutrition management.

Regarding forage crops, a cultivating system adaptive to climate change that can prevent yield decrease is being established, by cultivating multiple species and scattering the harvesting season. Development and dissemination of summer heat measures including manuring management techniques and varieties or breeding materials resistant to heat or wide range of maturation have been promoted. Also, measures against disease and pest have been taken, such as development and dissemination of resistant varieties and breeding materials.

Measures taken so far will be continuously promoted hereafter.

---

[9] A device made by combining a water-dampened cooling pad and a cooling fan, which obtains cooling effects by evaporatively cooling a greenhouse for agriculture

[10] A technology that efficiently uses thermal energy by a small amount of input energy

9.2.2.7 Plant Pests, Weeds, and Animal Infectious Diseases

It is important to understand appropriately the situations of the occurrence and damage of pests presented in Japan. For this reason, a pest forecasting program applicable to the specified pests[11] will be continuously implemented, and while investigating changes in the status of the occurrence and damage, such information will be transmitted for the purpose of timely and proper pest control. Moreover, in response to climate change, the specified pests for the pest forecasting program and the pest control system should be reviewed.

As for serious pests[12] that are not yet observed in Japan, or observed only in limited regions, import quarantines to prevent introduction of pests from overseas, domestic quarantines to prevent such pests from spreading in Japan, monitoring and controlling measures of the pests have been taken. In addition to that, further risk assessment of pests based on global information is to be implemented, and quarantine measures will be reviewed based on its results.

Regarding serious pests that are already observed in Japan, efforts will be taken accordingly to prevent them from spreading further, and to advance control techniques.

As for long-range migratory pest insects,[13] efforts are to be made in order to develop technology to monitor changes of migration information from overseas (i.e., the period of migration and the number of migrated insects) and technology to project the change of distribution areas in Japan (i.e., northward expansion of the overwintering area and the acceleration of occurring and migration).

Regarding diseases such as rice sheath blight and rice stripe, which is expected to increase in paddy fields, impacts on paddy rice yields are to be investigated, and technologies against such impacts are to be developed.

Regarding weeds, the risk of spoiled grains of soybeans due to increases in the amount of weeds remaining until the harvesting season will be assessed, and development of technology that can mitigate the damage will be promoted.

As for animal infectious diseases, the following efforts will be taken: selecting viral strains (viruses for manufacturing vaccines suitable for epidemic infectious diseases) as possible vaccine candidates against livestock infectious diseases carried by arthropods, examination of risk management including effective prevention measures, and investigation on migratory birds, which are known as a major transmitter of avian influenza.

---

[11] As specified by the Minister of Agriculture, Forestry and Fisheries in Article 22 of the Plant Protection Act (Act No. 151 of 1950), pests and diseases of which distribution in Japan is not limited to certain area, and they spread quickly and tend to do material harm to crops.

[12] Pests and diseases that possibly do material harm to useful plants if they spread within the country

[13] Insect pests that travel from several hundred kilometers to several thousand kilometers using not only their own flying capability, but also a large-scale weather phenomenon, including many serious insect pests in agriculture such as plant hoppers, aphids, and Noctuidae. Insect pests such as rice brown plant hopper (*Nilaparvata lugens*) and the white backed plant hopper (*Sogatella furcifera*) are known to migrate mainly to western Japan across from mainland China by a low-level jet stream that develops during the rainy season in Japan.

9.2.2.8 Water, Land, and Agricultural Infrastructure

"Global Warming Measures in Agricultural and Rural Development" was compiled, and investigations on measures concerning water, land, and agricultural infrastructure are being conducted. Development of technologies that contribute to project the impacts of global warming is promoted based on the technical development plan of agricultural and rural development.

Considering the impacts of high temperature and decline in snowmelt runoff volume that are expected in the future, agricultural water will be effectively secured and utilized by appropriately combining measures in hard (structural) and soft (nonstructural) aspects as follows: reducing the amount of irrigation water by automated water management and pipelining waterways and effective utilization of existing water resources by optimizing the operation of reservoirs and agricultural dams.

In order to respond to increasing torrential rainfall, efforts will be taken for disaster prevention in rural areas, as well as the maintenance and improvement of disaster reduction functions, by appropriately combining measures in hard (structural) and soft (nonstructural) aspects as follows: promoting prevention of farmland flooding by developing drainage pumping stations and drainage canals, clarifying facilities and regions highly vulnerable to inundation, conducting risk assessment such as formulating hazard maps, and promoting the development of a business continuity plan by facility managers. In doing so, measures will be effectively taken by utilizing existing facilities and connections of the local community.

Under current circumstances, climate change projections are highly uncertain, and there are no sufficient grounds for conducting specific examinations based on projected impacts. Thus, in light of new scientific findings, accompanied by the advancement of climate change research, medium- to long-term impacts will be projected and assessed.

If new scientific findings and climate models, improved assessment methods of impacts on basic agricultural infrastructure are developed so that the principle for facility maintenance based on projected impacts is clarified, desirable measures for such infrastructure will be reconsidered.

9.2.2.9 Securing Food and Feed Safety (Agricultural Products Including Cereal Grain and Cereal Products and Feeds)

Through continuous monitoring of the distribution of mycotoxin[14]-producing fungi in domestic field soil and mycotoxin contamination of domestic agricultural products and feeds, efforts will be made to ascertain the impacts of climate change. If increased mycotoxin contamination of agricultural products and feeds are likely to pose health problems for humans and livestock, technologies to reduce contamination will be developed and disseminated to producers of agricultural products and

[14] Substances that have harmful effects on humans and livestock among natural chemical substances made by mold

feeds. Mycotoxin contamination reduction measures will be regularly verified and reviewed in consideration of new findings.

9.2.3 Revising the Adaptation Plan

It is stated in the National Adaptation Plan (2018) to implement and formulate an assessment of climate change impacts in 2020 and, based on the results, status of each measures and global trends, to revise the Plan in 2021.

References

Central Environment Council: Report on impact assessment of climate change in Japan and future issues (Opinions) (March 2015) (Japanese) http://www.env.go.jp/press/upload/upfile/100480/27461.pdf

National Plan for Adaptation to the Impacts of Climate Change (Cabinet Decision on 27 November 2018) (Japanese) http://www.env.go.jp/earth/tekiou/tekioukeikaku.pdf

National Plan for Adaptation to the Impacts of Climate Change (Cabinet Decision on 27 November 2015) (English) http://www.env.go.jp/en/focus/docs/files/20151127-101.pdf

Chapter 10
Information Platform for Local Governments in Japan

Masashi Okada

Abstract In response to changing climate, Japan formulated its first National Adaptation Plan (NAP) in 2015. A portal site, A-PLAT, was funded by the Ministry of the Environment and developed by the National Institute for Environmental Studies, Japan, in 2016 to serve as an information platform to support effective local adaptation in Japan based on the NAP. A-PLAT consists of the following pages: (1) a basic explanation of climate change and adaptation, (2) information for specific stakeholders, (3) national and local policies, (4) archives of scientific reports, (5) guidelines for local government for their planning, and (6) webGIS displaying research outcomes that project future impacts on major indices in the NAP. Additionally, a local adaptation support tool will be implemented on A-PLAT within a few years. Since its launch, A-PLAT has played an important role in promoting adaptation strategies: local governments increasingly rely on A-PLAT for their planning, and their officers have contacted A-PLAT to seek assistance. As public interest in climate change grows, the government plans to enforce an act regarding adaptation in 2018. The ministry also plans to expand the function of A-PLAT to cover the Asia-Pacific region. It is evident that A-PLAT has been leading adaptation actions in Japan.

10.1 Introduction

The progress of global warming has already impacted various sectors throughout Japan. For the water resources sector, the variability in yearly total precipitation has been expanding, thereby increasing the risk of flood and drought (Japan Meteorological Agency (JMA) 2018). For the ecosystem sector, the cherry blossoms have been blooming earlier and maple leaves have been changing later in various locations of Japan with increasing temperatures (JMA 2018). It has also been

M. Okada (✉)
Center for Social and Environmental Systems Research, National Institute for Environmental Studies, Tsukuba, Ibaraki, Japan
e-mail: okada.masashi@nies.go.jp

T. Iizumi et al. (eds.), *Adaptation to Climate Change in Agriculture*,
https://doi.org/10.1007/978-981-13-9235-1_10

reported that the habitat range of *Cervus nippon* and hogs has been expanding as the area covered by snow changes (Japan Wildlife Research Center 2012), and the habitat area of *Papilio memnon* Linnaeus, which is widely distributed in warm areas, has been moving northward (Kitahara et al. 2001). In the ocean surrounding Japan, while northern species have been decreasing, southern species have been extending to the north due to increasing sea water temperature (Fisheries Agency of Japan (FAJ) 2017); coral whitening and seaweed bed loss have also been observed (FAJ 2015, 2017). The agricultural sector has also experienced various climate change impacts: immature rice grains with white portions due to hot summer (Ministry of Agriculture, Forestry and Fisheries (MAFF) 2018); the rind puffing of the *Citrus unshiu* due to high temperatures and excessive rain during the period from autumn to winter (MAFF 2017); and the poor coloring of grapes due to high temperatures during the period from summer to autumn (MAFF 2018). The human health sector has also seen some impacts: an increasing trend in the number of deaths resulting from heat stroke and an extending trend of the northernmost range for tiger mosquitoes (Ministry of the Environment Government of Japan (MoEJ) 2018a; Kobayashi et al. 2008). Because the mosquitoes are known as ones that transmits dengue, their expanding means to enhance potential risk of infection with dengue fever. The future progress of global warming is likely to cause more severe damage and to spread that damage to all sectors in all local areas throughout Japan.

Understanding a concept of climate change risk is crucial for effective adaptation action. In the Fifth Assessment Report of the IPCC Working Group II, the risk concept is represented as the probability of hazardous consequences from the interaction of hazards (potential natural or human-induced physical events or trends or their physical impact) with vulnerability (sensitivity or susceptibility to harm and a lack of capacity to cope and adapt) and exposure (the degree to which people and things could be adversely affected) (Oppenheimer et al. 2014). The hazards, vulnerability, and exposure have a very close relationship with specific natural and socio-economic characteristics in each local area and/or each sector. Therefore, it is necessary to develop legal and social systems fitted to local characteristics to develop effective adaptation measures to reduce our vulnerability to the harmful risks of mid- and long-term changes in climate and seize any potential beneficial opportunities associated with climate change.

In Japan, to promote comprehensive and systematic actions for climate change adaptation, the "National Plan for Adaptation to the Impacts of Climate Change" was approved in 2015 as Japan's first National Adaptation Plan (NAP) (MoEJ 2015). Subsequently, Japan's first national adaptation act, the "Climate Change Adaptation Act," was enacted in June 2018 and was enforced in December 2018 (MoEJ 2018b). A-PLAT was launched in 2016 as an information platform to support local adaptation in Japan according to NAP. A-PLAT, which was legally clarified by the enforcement of the act, is playing an increasingly important role in facilitating local adaptation. After briefly explaining NAP and the Adaptation Act, which contributed to the development of A-PLAT in the next section, A-PLAT and related activities will be introduced.

10.2 Legal Circumstance to Development of Information Platform

10.2.1 Climate Change Adaptation Act in Japan

The purpose of the Climate Change Adaptation Act enacted in June 2018 and enforced starting in December 2018 (MoEJ 2018b) is to promote climate change adaptation activities through the formulation of plans for climate change adaptation and the provision of information about climate change impacts, adaptation, and necessary measures, thereby contributing to the healthy and cultured living of the Japanese people both today and into the future. One important element of the act is the establishment of clear roles for national and local governments, the private sector, and citizens to promote climate change adaptation efforts. The national government plays a role by formulating and promoting policies and measures and establishing systems to collect, organize, and provide data about climate risks and adapting to those risks; this work contributes to the formulation of adaptation policies by local governments and the promotion of adaptation activities by the private sector and citizens. The role of local governments is to formulate and implement local adaptation policies incorporating local natural and socioeconomic characteristics while considering the NAP. And the local government has a role to promote adaptation practices for local private sector businesses and citizens. The private sector should work on adaptation practices in conjunction with its business activities, and it should cooperate with the adaptation policies of national and local governments. Citizens should enhance recognition and understanding of the importance of adaptation measures.

To address adaptation planning and impact assessment nationwide, the national government formulated the NAP to promote adaptation activities in each sector. The NAP executes risk management so that all adaptation processes can smoothly progress without requiring extensive reworking. The basic approach is an adaptive approach that cycles through conducting ongoing observation, monitoring, implementing regular assessments of climate change and its impacts, considering and implementing adaptation measures based on the results of the assessment, monitoring the conditions of progress, and revising as necessary. The NAP is then revised as required based on an impact assessment, but a routine assessment is implemented approximately every 5 years.

For adaptation planning in a local area, local governments should formulate their own Local Climate Change Adaptation Plan (LAP) by considering the local natural and socioeconomic characteristics alongside the NAP. Moreover, local governments are encouraged to create a local climate change adaptation center as a hub for collecting, organizing, analyzing, and providing local data on climate risks and adaptation and for providing technical advice to local stakeholders such as the government, the private sector, environmental associations, and citizens. To engage in the necessary discussion around promoting local adaptation, local stakeholders can organize local climate change councils in cooperation with local MoEJ branches, local governments, and local climate change adaptation centers and businesses. To ensure that local adaptation measures proceed appropriately, the national government provides necessary information and related support through A-PLAT.

10.2.2 National Adaptation Plan of Japan

The NAP of Japan formulated in 2015 states its vision: to build a secure, safe, and sustainable society that can minimize or avoid impacts from climate change risks on people's lives, bodies and property, national and local economies, and the natural environment and that can quickly recover, by adjusting social and natural conditions through appropriate adaptation measures, whatever impacts occur due to climate change (MoEJ 2015). To systematically and comprehensively promote adaptation, the following five basic strategies are established:

1. Mainstream adaptation into government policies and measures by building resilience, considering the uncertainties of climate risks, creating synergies through adaptation, and researching, developing and disseminating adaptation technologies.
2. Continuously enhance scientific findings with ongoing observation, monitoring, projection, and assessment and the promotion of studies and research.
3. Promote understanding and cooperation through sharing and providing accurate and comprehensible information about climate risks and related topics.
4. Promote local adaptation efforts by cooperating with local governments for climate change impact assessments, formulating adaptation plans, and raising awareness.
5. Promote international cooperation and contribute to developing countries by providing assistance in formulating adaptation plans and implementing adaptation measures by making use of human resource development and Japanese science and technology.

The contents of the NAP consider seven sectors: agriculture, forest/forestry and fisheries, water environment/water resources, natural ecosystems, natural disasters/coastal areas, human health, industrial and economic activity, the life of the citizenry and urban life. The NAP states that in regard to concrete adaptation measures, it is important to base them on factors such as the local socioeconomic situation in Japan, local and sectoral characteristics, and the extent of climate change impacts.

10.3 Information Platform to Facilitate Implementation Adaptation in Japan

One of the basic strategies behind the NAP is to enhance the understanding and cooperation of all stakeholders by organizing and sharing climate risk and other information, building bridges with policy making, promoting awareness, and

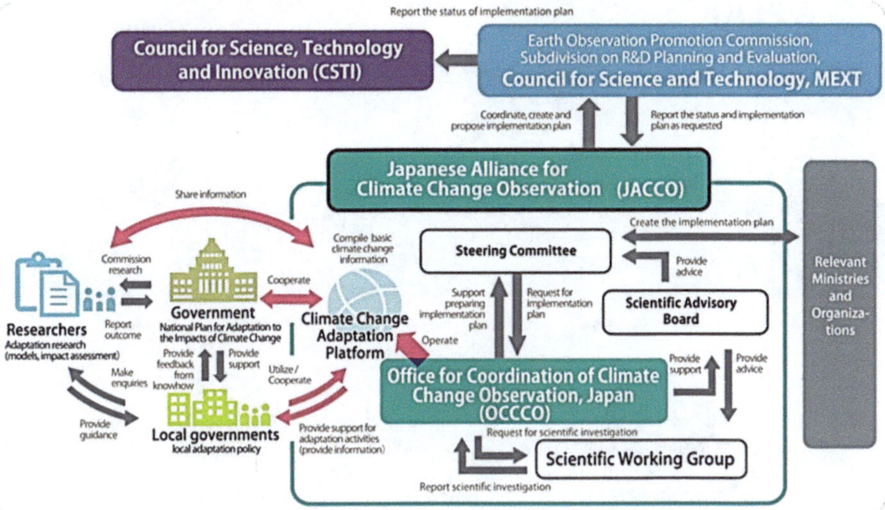

Fig. 10.1 Organizational structure surrounding the Climate Change Adaptation Information Platform (A-PLAT) (This figure was provided by Ms. Fukumura)

enhancing human resources (see Sect. 10.2; MoEJ 2015). To implement this strategy, the Climate Change Adaptation Information Platform (A-PLAT) was launched in August 2016 as the core of an information infrastructure designed to assist in the adaptation actions of various stakeholders such as local governments, businesses, and citizens (Fig. 10.1). A-PLAT was developed and has been operated by the National Institute for Environmental Studies (NIES), Japan, and is funded by the MoEJ in cooperation with the related government ministries and agencies. A-PLAT is a one-stop online resource for information on adaptation to climate change impacts for various stakeholders in Japan, and it has been collecting, organizing, analyzing, and providing information related to climate risks and adaptation and providing technical support to local stakeholders. In addition, A-PLAT aims to disseminate knowledge of adaptation to the public to enhance understanding and recognition. The A-PLAT portal can be accessed in Japanese (Fig. 10.2) or in English (NIES 2018). The principal pages are the following:

1. A basic explanation of climate change and adaptation
2. Information about adaptation policies, plans, and related projects of the national government
3. Information about adaptation policies, plans, and practices of local governments
4. Information about climate risk management and adaptation for businesses in the private sector
5. A webGIS displaying research outcomes that project future climate and impacts on major indices in the NAP
6. An introduction to adaptation measures for citizens

Fig. 10.2 Front page of the A-PLAT portal site

Climate projections and quantitative climate change impact assessments are vital information required by local governments and other stakeholders to prepare adaptation plans. A-PLAT has been collecting results from the latest research projects to include in their impact assessment. However, it is not easy for nonexperts to use and handle these research data due to the huge volumes of data and specific file formats. To overcome this issue, A-PLAT adopts a webGIS mapping tool to provide a user-friendly environment that allows everyone to easily view and download interesting data. As demonstrated in Fig. 10.3, by simply clicking the prefectural name on the page and then selecting items of interest such as sector, index, General Circulation Model (GCM), greenhouse gas (GHG) scenarios, and target period, the user can easily draw an objective map and graph showing climate projections and quantitative impact assessments. In November 2018, A-PLAT implemented the most recent research project on impact assessment in Japan (Comprehensive Study on Impact Assessment and Adaptation for Climate Change (S-8 Strategic Research Project)) developed by the MoEJ (S-8 Climate Change Impact and Adaptation Research Project Team 2014) as well as the latest climate scenarios developed by the JMA (JMA 2017). Users can view and use the research outcomes from these climate projections and impact assessments for five sectors (agriculture, water resources,

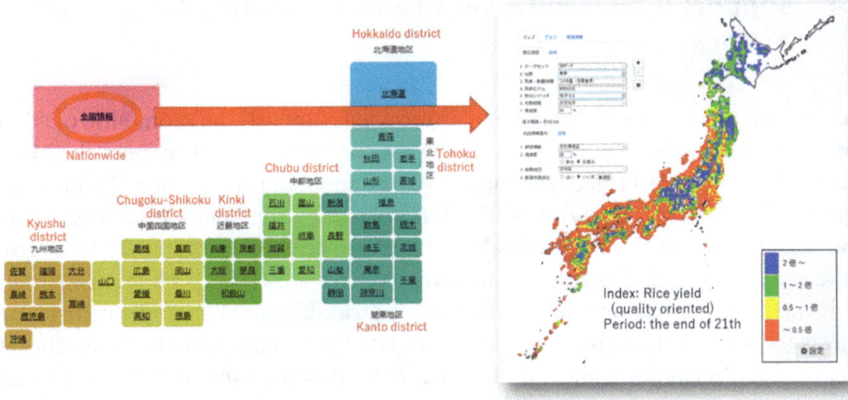

Fig. 10.3 Demonstration of climate change impact viewer in A-PLAT portal site

natural ecosystems, natural hazards, and health) nationwide and for each prefecture, which cover 17 items of assessment (e.g., rice yield for agricultural sector), 4 GCMs (MIROC5 (Watanabe et al. 2010), MRI-CGCM3.0 (Yukimoto et al. 2011, 2012), GFDL CM3 (Delworth et al. 2006; Donner et al. 2011), Had GEM2-ES (Collins et al. 2011; Martin et al. 2011)), 3 levels (2.6, 4.5, and 8.5) of Representative Concentration Pathways (RCP, which is GHG concentration trajectory adopt by the IPCC for its fifth Assessment Report), and 3 periods. MIROC5 and MRI-CGCM3.0, which were developed by Japanese research institutes, provide a highly reliable projection of climate, monsoon, and the baiu front (seasonal rain front) on Asia including Japan. The four GCMs were selected to cover variation range among projection results from all GCMs in air temperature and precipitation around Japan (S-8 Climate Change Impact and Adaptation Research Project Team 2014).

A-PLAT collects and disseminates the latest information and news about adaptation based on the work of each stakeholder. For the national government, A-PLAT introduces the NAP, the Adaptation Act, and related efforts and summarizes national research projects on climate change impacts and adaptation. For local governments, A-PLAT lists the LAPs and the related efforts by each local government, provides guidelines for municipal adaptation planning, and introduces the consortium project for local adaptation (see Sect. 10.4 for details). For the private sector, A-PLAT provides information and examples of climate risk management and adaptation by private companies. For citizens, A-PLAT provides a straightforward explanation of climate risk and adaptation with friendly illustrations that can be easily understood by everyone, young and old alike (see Sect. 10.6 for the related explanation). The examples of risks affecting citizens show heatstroke caused by increasing temperature and effect of natural disasters from extreme weather events such as localized heavy rain.

10.4 Support for Local Governments Promoting Adaptation Actions

As explained in Sect. 10.2, the national government supports the promotion of adaptation in the local areas (MoEJ 2015, 2018b). As one government support, the MoEJ initiated the "Support Project for Impact Assessment and Adaptation Planning for Climate Change" in 2015 with A-PLAT (Global Environment Committee of the Central Environment Council of the MoEJ 2017). The MoEJ selected 11 local governments (Fukushima, Saitama, Kanagawa, Mie, Shiga, Hyogo, Ehime, Nagasaki, and Kumamoto prefectures and Sendai and Kawasaki cities) to implement model projects to assist in arranging information about climate change and its impacts and in formulating adaptation plans in the local area. Based on the model projects, the respective local governments have been formulating their own adaptation strategies. From the knowledge obtained from the model projects, the MoEJ established guidelines published in the "Municipal Adaptation to the Impacts of Climate Change – Planning Guidelines" in 2016 (MoEJ 2016). The guidelines are conceptual and provide a stepwise procedure for local governments to formulate an adaptation plan. For detail, refer to MoEJ (2016), but the basic steps are the following: Step 1 is to consider the short-term goal and the process; Step 2 is to establish an institutional framework to promote adaptation; Steps 3 and 4 are to compile ongoing and projected climate data and impacts; Step 5 is to compile information on existing measures; Step 6 is to conduct a climate change impact assessment; Step 7 is to formulate adaptation plans; and Step 8 is to share and provide information with the community.

Moreover, bolstered by the success of the A-PLAT, the MoEJ started a consortium project for local adaptation to climate change with related ministerial offices in 2017 closely supported by A-PLAT as its secretariat (Global Environment Committee of the Central Environment Council of the MoEJ 2017). The main purpose of the consortium is to promote the local assessment of climate change and its impacts and the formulation and implementation of the respective local adaptation plans. The consortium has been conducted in six districts in Japan (Hokkaido-Tohoku, Kanto, Chubu, Kinki, Chugoku-Shikoku, and Kyusyu-Okinawa). Each district not only collects and organizes existing information about climate risk but also assesses the local impact independently in cooperation with related persons in national ministerial offices, local governments, and research institutes. The project is expected to develop a local framework that gives local governments the ability to execute an impact assessment and formulate adaptation plans individually within their respective districts.

The content of the model projects, guidelines, consortium projects, and other information discussed above are summarized and introduced in the A-PLAT site.

10.5 Examples of Adaptation Plans in Local Governments

Hazards, vulnerabilities, and exposures to climate change risks vary substantially with local natural and socioeconomic characteristics. Therefore, each local government needs to consider its own adaptation measures proactively and address them carefully. A local adaptation plan should include the following:

- Formulation of a unique plan for local adaptation
- Consideration of local-specific climate change projections
- Consideration of local-specific impacts from climate change from various viewpoints such as seriousness and urgency
- Formulation of additional or a new adaptation plans alongside existing policies
- Consideration of adaptation measures involving all local stakeholders, meaning not only the public sector but also the private sector and citizens
- Consideration of citizens' attitudes toward adaptation
- Establishment of indices and numerical targets to conduct ex post evaluation
- Arrangement of a promotion system by the public and private sectors

The material below introduces an advanced effort toward adaptation by local governments, taking Tokushima prefecture and Saitama prefecture as examples.

In Tokushima prefecture, which is located in the Shikoku district in western Japan, the local government became the first government entity in Japan to enact an ordinance that introduces a full range of adaptation measures (Tokushima prefecture 2016). The LAP in Tokushima prefecture was formulated in six sectors: preservation of prefectural land, natural ecosystems, water environment/water resources, human health, industrial and economic activities, and agriculture and fisheries. While the LAP can cover the sectors formulated in the NAP (agriculture, forest/forestry and fisheries, water environment/water resources, natural ecosystems, natural disasters/coastal areas, human health, industrial and economic activities, and life of citizenry/urban life), the prefectural government determined its own unique sectors and an adaptation plan for each sector based on local natural and socioeconomic characteristics. The agricultural sector in Tokushima prefecture has unique characteristics: various agricultural products are produced based on the local natural environment, and the products hold a high market share in the Kansai district, which is one of the major consumption areas in Japan. Therefore, agriculture is an important industry in Tokushima prefecture. To establish a stable agricultural production system and create new local agricultural brands under new climate conditions, the Department of Agriculture, Forestry and Fisheries of Tokushima prefecture is working on the development of cultivation and feeding management techniques and crop varieties adaptive to high-temperature conditions and the introduction of highly valued southern fruits in association with various relevant research institutes including a local university, Tokushima University. They established the detailed targets, one of which is that they develop high-temperature-resistant new varieties from 11 varieties in 2013 to 15 in 2018. According to the progress as of 2017, the number of the varieties achieved to 14. The other sectors aim to realize a safe and secure society such that climate change risks are reduced by considering local characteristics and the local cities, towns and villages are revived in some areas by effectively utilizing the positive impacts of climate change. Tokushima prefecture formulated its own adaptation plans based on local characteristics, as its topography is steep with vulnerable soil, it is frequently hit by typhoons, and its aging population ratio is greater than the national average. The LAP also makes ambitious efforts to not only show the vision of the adaptation plan but also set indices and numerical targets; the NAP does not set numerical targets.

Saitama prefecture, which adjoins the north side of the metropolis of Tokyo, developed a plan for prefectural global warming countermeasures called "Stop Global Warming: Saitama Navigation 2050" in 2009 (Saitama prefecture 2015). Saitama had been promoting prefectural adaptation activities more than 5 years prior to the formulation of the NAP in 2015. Based on its accumulated experience, the prefecture summarized how to efficiently promote policies that contribute to climate change adaptation. To proceed with adaptation measures involving the entire prefecture, close cooperation is necessary among all stakeholders, such as governments, the private sector, citizens, and environmental associations, with each performing its respective role and fulfilling its respective responsibilities. Saitama prefecture has clearly defined the role of each stakeholder. For example, the prefectural government should develop plans for all stakeholders to proactively take measures to prevent global warming and adapt to climate risks. The governments of cities, towns, and villages should proactively take adaptation measures closely tied to local communities and based on the local natural and social conditions, as the local governments are closest to citizens. Citizens should take an environmental education and activities. The private sector should take measures that consider the environment, such as providing environmental education for employees. Environmental associations are expected to propose adaptation measures to governments, citizens, and the private sector utilizing their extensive experience and expert knowledge. To promote adaptation measures across the entire prefecture, Saitama prefecture took the initiative by constructing a comprehensive framework including cooperation with governments, the private sector, citizens, environmental associations, and related organizations.

On July 23, Kumagaya City in Saitama prefecture reached 41.1 °C, the highest temperature ever recorded in Japan. As one of the causes, global warming may well have been incriminated. Therefore, Saitama prefecture has been working to promote adaptation actions to hot summer environment with the prefectural research institute, Center for Environmental Science in Saitama. The prefecture has been calling for cooperation in setting a resting place for prevention of heat stroke on local private facilities such as convenience store as well as public facilities since 2011. In 2017, the cooperated facilities reached 6604. As the other adaptation action, in order to reduce the heat stress for users of a large sports recreation park in the prefecture, the prefecture assessed the reduction effect of heat temperature condition due to change in pattern of tree planting and introducing heat insulation pavement by using a computer simulation. The assessment results used to optimize the policies on the park.

The projections of climate change and its impact assessments are uncertain, although the uncertainties are decreasing through the efforts of researchers. Under progressing global warming, stakeholders need to use the current research findings to examine future impacts and take adaptation measures. Saitama prefecture has prepared multiple actions based on the various levels of impacts and their uncertainty alongside existing policies, and it is gradually implementing projects according to the monitoring results through a mechanism called the "promotion of adaptation with flexibility." Along with this mechanism, Saitama prefecture has been working to enhance the effectiveness with which it is promoting adaptation while also cooperating with governments internal and external to the prefecture, the private sector, citizens, and environmental associations.

Generally, a LAP is created based on the local information of climate change and its impact assessments. However, there are few local governments having enough such the information accumulated by their own work. The most local governments have a common problem that collecting and compiling the information from their outside takes a lot of time and efforts. A-PLAT has been collecting and organizing the latest scientific information of climate change and its impact assessments in all of Japan as well as the most recent adaptation plans and related information from all local governments in Japan and then disseminating these materials sequentially from the A-PLAT portal site. In addition, A-PLAT interviews those involved and reports detailed adaptation actions when interesting cases emerge such as the above examples of Tokushima and Saitama prefectures. Therefore, because the A-PLAT provides local data on climate change and its impact assessment together with the scientific and technical advices and the related materials, adaptation plan developers in local governments can systematically collect the related information to consider their own plan.

10.6 Promoting Adaptation Action by Citizens

To promote adaptation measures for an entire local area, it is necessary for citizens and communities to enhance their understanding and recognition of climate risks and adaptation. There are many adaptation measures related to actions undertaken by citizens and communities, such as preventive measures for heatstroke, practices to save water, and the confirmation of local hazard maps, which can contribute to securing lifestyles. The result of a public opinion survey on measures against global warming (2016 survey) shows that the Japanese people are highly interested in climate change impacts (Cabinet Office 2016). In particular, 63.1% fear an increase in the risk of natural hazards, including floods, tidal waves, and high waves, and 57.7% of them are worried about a decline in the yield and quality of crops. However, the ratio of Japanese people who recognized the term "adaptation," meaning a measure to reduce negative impact due to climate change, is only approximately half (47.5%), while relatively few knew any details about adaptation (only 4.3%). The survey concluded that although many Japanese people know about climate change impact, their recognition of adaptation is not as high.

In each local area, there are citizens, communities, NGOs, and related associations that are working on global warming issues that systematically carry out environmental prevention activities. Most of these have local-specific expert knowledge about climate risks and adaptation measures based on local natural and socioeconomic characteristics. Such citizens and communities are expected to contribute to better understanding and recognition among other groups that are not familiar with climate impacts and adaptation. To assist in creating bridges, local governments and A-PLAT have contributed to developing a framework to promote adaptation measures across an entire local area with all stakeholders and involving all citizens living in the local area.

10.7 Toward Developing an Information Platform for Countries in the Asia-Pacific Region

There are some international networks to promote adaptation actions, one major network being the Global Adaptation Network (GAN). Founded in 2010 by the United Nations, the overall objective of the GAN is to help build resilience against climate change through the mobilization of adaptation knowledge. The first regional network launched under the GAN is the Asia Pacific Adaptation Network (APAN). It was established in October 2009 in Bangkok by the Prime Minister of Thailand with support from the Japanese and Swedish governments and the Asian Development Bank. The APAN aims to promote the mobilization of knowledge, best practices, and technology to improve the resilience and sustainability of social systems, ecosystems, and economies to climate change by supporting capacity building for the adaptation and design of policies and planning in Asia and the Pacific regions.

Japan will also strengthen international cooperation, as described in the NAP and the Climate Change Adaptation Act, to facilitate partnerships and cooperation for climate-resilient and sustainable development in the countries in the Asia-Pacific region. Japan responded to the Paris Agreement at the 22nd session of the Conference of the Parties (COP23) to the UN Framework Convention on Climate Change in 2016 by declaring the launch of the "Asia-Pacific Adaptation Information Platform (AP-PLAT)" by 2020. The overall objective of the AP-PLAT is to support countries in the Asia-Pacific region by assisting with climate change impact assessment and the development of national adaptation plans that facilitate adaptation to the changing climate. The main tasks of AP-PLAT are the following:

1. Arranging and disseminating information about adaptation plans and measures taken by countries in the Asia-Pacific region

2. Collecting, organizing, and disseminating scientific information about climate change projections and impact assessments and developing a dataset for the projection of climate change impacts in the region through bilateral and intensive research studies
3. Developing supporting toolkits for officials and stakeholders engaged in adaptation planning
4. Introducing capacity-building projects and bilateral environmental partnerships for climate change impact assessment and adaptation planning
5. Collecting and disseminating the information about adaptation for businesses and financial information to promote national adaptation measures

Although the portal site of the AP-PLAT is still under development by NIES in November 2018, the current version can be accessed on http://www.adaptation-platform.nies.go.jp/en/ap-plat/index.html (Fig. 10.4). The AP-PLAT site provides six types of content: (1) an introduction of basic information about climate change and adaptation, (2) plans and implementations (e.g., aggregation of NAPs), (3) tools and guidelines (e.g., webGIS on climate impact assessment), (4) financial information on adaptation project implementation (e.g., Green Climate Fund, Global Environment Facility, Adaptation Fund, Climate Investment Fund), (5) prototype of adaptation business, and (6) news. One of the important functions on the site is the

Fig. 10.4 Frontpage of the AP-PLAT

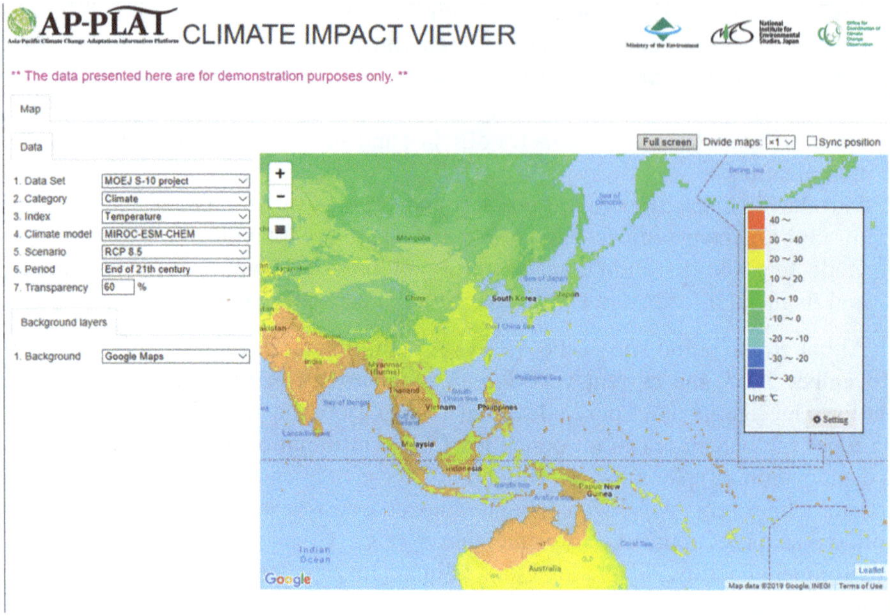

Fig. 10.5 Demonstration of climate impact viewer in the AP-PLAT

climate impact viewer (Fig. 10.5). By adopting the webGIS mapping tool, AP-PLAT provides a user-friendly environment where everyone can easily see research outcomes for climate change projections and impact assessments of the Asia-Pacific regions, such as A-PLAT (See Sect. 10.3). The climate impact viewer shows climate change impact assessment results based on Integrated Climate Assessment-Risks, Uncertainties and Society (ICA-RUS), Comprehensive Research on the Development of Global Climate Change Risk Management Strategies (S-10 Strategic Research Project) supported by the MoEJ (Emori et al. 2018). In January 2019, the viewer implemented future climate projection and the impact assessment results on three sectors (terrestrial ecosystems, water resources, and health) based on the climate scenario from MIROC ESM-CHEM (Watanabe et al. 2011) with four levels (2.6, 4.5, 6.0, and 8.5) of RCP. The number of the climate model and the impact assessment items will be increased in the future.

On the pages of (2) Plans and Implementations, AP-PLAT shares the adaptation plans for the Asia-Pacific countries laid out using inputs from Nationally Determined Contributions (NDCs), National Adaptation Plans (NAP), National Adaptation Programmes of Action (NAPA), and Technology Needs Assessments (TNA) in order to provide the country needs on adaptation. The NDCs are a core element of the Paris Agreement and are critical for the achievement of these long-term goals. Specifically, the NDCs embody the post-2020 climate actions or efforts made by each country to adapt to the impacts of climate change. The Paris Agreement (in Article 4, paragraph 2) requires each country to outline and communicate these efforts. The TNAs are a long-standing process under the UNFCCC. Since 2001, more than 80 developing countries have undertaken them to assess their technological needs and address climate change. The Green Climate Fund (GCF) secretariat identifies developing countries' NDCs and TNAs as important reference points for GCF programming. Sharing the adaptation plans is helpful for adaptation project developers.

And AP-PLAT shows also international cooperation on climate change impact assessment and adaptation measure by each country/organization on the pages of (2) Plans and Implementations. The MoEJ has been establishing a consortium including research institutes and consultants in host countries and performing a survey of needs for the formulation of adaptation plans and an assessment of climate change impacts under bilateral environmental partnerships since 2015. The bilateral environmental partnership has been implemented in Indonesia (supporting climate change impact assessment for the formulation of a local adaptation plan in Indonesia), in Mongolia (supporting climate change impact assessment in Mongolia), in the Asia-Pacific region (investigating and examining to develop human resources to promote adaptation initiatives), in the Pacific region (a survey for methods to identify the needs of climate change impact assessments or detailed policy and a survey on cooperation in the implementation of climate change impact assessments and detailed policy in the Pacific region), and in island countries (supporting climate change impact assessments in the Pacific region). The MoEJ, with the cooperation of JMA, created a climate change projection dataset in countries of the Asia-Pacific region for contributing to promoting the regional climate change adaptation planning. The dataset includes outputs of global climate model and regional climate model. AP-PLAT introduces the reports from the bilateral environmental partnerships and the climate change projection dataset on the site.

AP-PLAT also collects information on financial and technological support mechanisms such as the Green Climate Fund (GCF), the Global Environment Facility (GEF), Adaptation Fund (AF), Climate Investment Fund (GCF), and the Climate Technology Center and Network (CTCN). The information is shared on the pages of (4) Financial information on adaptation project implementation. For developing countries, these mechanisms for adaptation are a very important support for introducing adaptation.

There are similar adaptation platforms other than AP-PLAT globally. Examples of platforms include the Climate Change Knowledge Portal (CCKP) by the World Bank (World Bank 2019), the European Climate Adaptation Platform (Climate-ADAPT) by the European Commission (European Commission 2019), The UK Climate Impacts Programme (UKCIP) by the United Kingdom (the United Kingdom 2019), and the IPCC Data Distribution Center by the IPCC Task Group on Data and Scenario Support for Impact and Climate Analysis (TGICA) (TGICA 2019). Climate-ADAPT is a desirable portal website for its relevant, reliable, timely and easy accessible contents, and UKCIP is a one for its administration that enables its stable operation to support continuous adaptation actions. The portal sites of A-PLAT and AP-PLAT are developed by adopting the content structure of Climate-ADAPT and the operation system of UKCIP and have been continued with those improvements. Compared to the other platforms, AP-PLAT plays a leading role to support efforts on climate change adaptation, especially for countries in the Asia-Pacific region, through the information distribution and the technical advices in cooperation with the national and local governments and their research institutes on the region. There are some international networks to promote adaptation, including GCEA (Global Center of Excellence on Climate Adaptation), which aims to provide expertise and advice to countries, communities, and companies on designing, financing, implementing, and monitoring adaptation actions; CLICC (Country Level Impacts of Climate Change), which aims to develop methodologies to help countries present climate change impact assessments in a consistent manner; as well as GAN and APAN. AP-PLAT has been working to support countries in the Asia-Pacific region in climate adaptation in close cooperation with other adaptation platforms and international networks.

10.8 Conclusion

The National Adaptation Plan (NAP) in Japan is expected to facilitate the promotion of local adaptation involving all stakeholders such as local governments, the private sector, and citizens in Japan. Based on the NAP, the A-PLAT portal site, an information tool that effectively assists local adaptation in Japan, was launched in 2016. Since its launch, A-PLAT has played a pivotal role in promoting adaptation measures for all stakeholders such as governments, the private sector, citizens, and communities in Japan. Local governments contact A-PLAT to seek assistance, ask for data to consider future impacts in their local area, request a scientific review of their newly created plan before its publication, etc. Bolstered by the success of A-PLAT, the ministry has started a consortium project to assess future impacts in local areas with A-PLAT. As public interest in climate change is growing, the national government enforced the Climate Change Adaptation Act in 2018. The government will also establish a portal site AP-PLAT by 2020 for communication and interaction to facilitate partnerships and collaboration in climate-resilient and sustainable development in the Asia-Pacific region. Now, Japan is in the very important phase of taking steps toward adaptation. By developing the information platforms, A-PLAT and AP-PLAT, we have facilitated adaptation actions in local areas of Japan and in countries of the Asia-Pacific regions.

Acknowledgments I thank Dr. Hijioka and Ms. Fukumura at NIES for providing reference materials in order to prepare this text.

References

Cabinet Office, Government of Japan (CAO) (2016) Public opinion survey on measures against global warming (2016 survey). CAO, Tokyo. Available at https://survey.gov-online.go.jp/h28/h28-ondanka/index.html in Japanese. Accessed 14 Oct 2018

Collins WJ, Bellouin N, Doutriaux-Boucher M et al (2011) Development and evaluation of an Earth-System model – HadGEM2. Geosci Model Dev 4:1051–1075

Delworth TL, Broccoli AJ, Rosati A et al (2006) GFDL's CM2 global coupled climate models. Part I: formulation and simulation characteristics. J Clim 19:643–674

Donner LJ, Wyman BL, Hemler RS et al (2011) The dynamical core, physical parameterizations, and basic simulation characteristics of the atmospheric component AM3 of the GFDL Global Coupled Model CM3. J Clim 24:3484–3519

Emori S, Takahashi K, Yamagata Y et al (2018) Risk implications of long-term global climate goals: overall conclusions of the ICA-RUS project. Sustain Sci 13:279–289

European Commission (2019) The European Climate Adaptation Platform (Climate-ADAPT). Available at https://climate-adapt.eea.europa.eu/. Accessed 21 May 2019

Fisheries Agency of Japan (FAJ) (2015) Coral reef conservation activity guidelines, p 69. Fisheries Agency of Japan, Tokyo. Available at http://www.jfa.maff.go.jp/j/gyoko_gyozyo/g_hourei/index.html#sango_hozen (in Japanese). Accessed 22 Jan 2019

Fisheries Agency of Japan (FAJ) (2017) Guidelines for climate change adaptation policy on development of fishing grounds, p 112. Fisheries Agency of Japan, Tokyo. Available at http://www.jfa.maff.go.jp/j/gyoko_gyozyo/g_hourei/attach/pdf/index-42.pdf (in Japanese). Accessed 22 Jan 2019

Global Environment Committee of the Central Environment Council of the Ministry of the Environment of Japan (MoEJ) (2017) Strategies on scientific knowledge's and information about climate-related risks for promotion of adaptation measures to climate change (interim summary). MoEJ, Tokyo. Available at http://www.env.go.jp/press/files/jp/105151.pdf (in Japanese). Accessed 13 Oct 2018

Japan Meteorological Agency (JMA) (2017) Global warming tukushmprojection volume 9, JMA, Tokyo, p 41. Available at https://www.data.jma.go.jp/cpdinfo/GWP/Vol9/pdf/all.pdf in Japanese. Accessed 29 Nov 2018

Japan Meteorological Agency (JMA) (2018) Climate change monitoring report 2017. JMA, Tokyo, p 87. Available at https://www.jma.go.jp/jma/en/NMHS/ccmr/ccmr2017_high.pdf. Accessed 22 Jan 2019

Japan Wildlife Research Center (JWRC) (2012) Investigation/evaluation report for mapping of Japan biodiversity outlook. JWRC, Tokyo, p 253. Available at http://www.biodic.go.jp/biodiversity/activity/policy/map/files/h23report_all.pdf (in Japanese). Accessed 22 Jan 2019

Kitahara M, Iriki M, Shimizu G (2001) On the relationship between the northward distributional expansion of the great Mormon butterfly, *Papilio memnon* Linnaeus, and climatic warming in Japan. Trans lepid Soc Jpn 52:253–264. (Japanese with English abstract)

Kobayashi M, Komagata O, Nihei N (2008) Global warming and vector-borne infectious diseases. J Disast Res 3:105–112

Martin GM, Bellouin N, Collins WJ et al (2011) The HadGEM2 family of Met Office Unified Model climate configurations. Geophys Model Dev 4:723–757

Ministry of Agriculture, Forestry and Fisheries (MAFF) (2017) Climate change impact investigation report 2016. MAFF, Tokyo, p 52. Available at http://www.maff.go.jp/j/seisan/kankyo/ondanka/attach/pdf/report-30.pdf (in Japanese). Accessed 22 Jan 2019

Ministry of Agriculture, Forestry and Fisheries (MAFF) (2018) Climate change impact investigation report 2017. MAFF, Tokyo, p 67. Available at http://www.maff.go.jp/j/seisan/kankyo/ondanka/attach/pdf/report-39.pdf (in Japanese). Accessed 22 Jan 2019

Ministry of the Environment Government of Japan (MoEJ) (2015) National plan for adaptation to the impacts of climate change. MoEJ, Tokyo, p 103. Available at http://www.env.go.jp/en/earth/cc/npa.html. Accessed 13 Oct 2018

Ministry of the Environment Government of Japan (MoEJ) (2016) Municipal adaptation to the impacts of climate change -planning guidelines. MoEJ, Tokyo, p 144. Available at http://www.adaptation-platform.nies.go.jp/lets/guideline_H28_08_env.pdf (in Japanese). Accessed 13 Oct 2018

Ministry of the Environment Government of Japan (MoEJ) (2018a) Environmental health manual on heat strokes. MoEJ, Tokyo, p 76. Available at http://www.wbgt.env.go.jp/heatillness_manual.php (in Japanese). Accessed 23 Jan 2019

Ministry of the Environment Government of Japan (MoEJ) (2018b) Climate change adaptation act. MoEJ, Tokyo, p 14. Available at http://www.env.go.jp/earth/tekiou/tekiouhou_jyoubun_r1.pdf (in Japanese). Accessed 12 Nov 2018

National Institute for Environmental Studies (NIES) (2018) A-PLAT portal site. NIES, Tsukuba. Available at http://www.adaptation-platform.nies.go.jp/ in Japanese and http://www.adaptation-platform.nies.go.jp/en/index.html (in English). Accessed 13 Oct, 2018

Oppenheimer M, Campos M, Warren R et al (2014) Chapter 19 emergent risks and key vulnerabilities. In: Field CB, Barros VR, Dokken DJ et al (eds) Climate change 2014: impacts, adaptation, and vulnerability. Part A: global and sectoral aspects. Contribution of working group II to the fifth assessment report of the Intergovernmental Panel on Climate Change. Cambridge University Press, Cambridge/New York, pp 1039–1099

S-8 Climate Change Impact and Adaptation Research Project Team (2014) Climate change "Impacts on Japan" -comprehensive impact assessment and adaptation measures based on new scenarios. Institute for Global Change Adaptation Science, Ibaraki University, Mito, Japan and National Institute for Environmental Studies, Tsukuba, p 46. Available at http://www.nies.go.jp/s8_project/symposium/20141110_s8english.pdf. Accessed 14 Oct 2018

Saitama prefecture (2015) Stop warming: Saitama navigation 2050 (updated version). Saitama prefectural office, Saitama, p 89. Available at https://www.pref.saitama.lg.jp/a0502/ontaikeikaku.html in Japanese. Accessed 14 Oct 2018

Task Group on Data and Scenario Support for Impact and Climate Analysis (TGICA) (2019) the IPCC Data Distribution Centre. Available at https://www.ipcc-data.org/index.html. Accessed 21 May, 2019

The United Kingdom (UK) (2019) The UK Climate Impacts Programme (UKCIP). Available at https://www.ukcip.org.uk/. Accessed 21 May, 2019

Tokushima Prefecture (2016) Climate change adaptation strategies in Tokushima prefecture. Tokushima prefectural office, Tokushima, p 56. Available at https://www.pref.tokushima.lg.jp/ippannokata/kurashi/shizen/2016110800025 (in Japanese). Accessed 14 Oct 2018

Watanabe M, Suzuki T, O'ishi R et al (2010) Improved climate simulation by MIROC5: mean states, variability, and climate sensitivity. J Clim 23:6312–6335

Watanabe M, Hajima T, Sudo K et al (2011) MIROC-ESM 2010: model description and basic results of CMIP5-20c3m experiments. Geosci Model Dev 4:845–872

World Bank (2019) Climate Change Knowledge Portal (CCKP). Available at https://climateknowledgeportal.worldbank.org/. Accessed 21 May 2019

Yukimoto S, Yoshimura H, Hosaka M et al (2011) Meteorological Research Institute-Earth System Model v1(MRI-ESM 1)—model description, Technical report of MRI no. 64. MRI, Tsukuba, p 88

Yukimoto S, Adachi Y, Hosaka M et al (2012) A new global climate model of the Meteorological Research Institute: MRI-CGCM3–Model description and basic performance. J Meteorol Soc Jpn 90A:23–64

Chapter 11
On Promoting Policy-Science Dialogue for Adaptation Planning in Agricultural Sector

Mariko Fujisawa and Hideki Kanamaru

Abstract Climate change poses considerable risks to agriculture and food security especially in developing countries. Taking urgent adaptation action is required, and the global agendas to support the countries are in place. An important element for adaptation planning is to build robust evidence about past and future climate risks and vulnerabilities and to identify and appraise options of adaptation practices.

However, many impact assessments in agriculture focus either on small-scale area, such as farm level, or global scale, and there is lack of information at national and sub-national level to address the climate risks for making adaptation policies. In many developing countries, the capacity to conduct assessments is also weak. In an attempt to fill these knowledge and capacity gaps, FAO has developed an integrated assessment tool MOSAICC and implemented in eight countries. It packages multiple models from different disciplines into one system to promote collaboration of an interdisciplinary group of subject experts and policy makers. Thus, we tried to promote a dialogue between science and policy. There are high expectations for science community to translate scientific research into practical risk and vulnerability assessment for decision-making at national and sub-national levels.

11.1 Introduction

Agriculture is vulnerable to climate change. 2.5 billion small-scale farmers in developing countries are most vulnerable to climate change and their food security is at risk. According to the latest UN report, the number of food insecure people in the world have declined from 945 million in 2005 to 784 million in 2014; however, the declining trend reversed since 2014, and the number of undernourished people reached an estimated 821 million in 2017 (FAO 2018).

M. Fujisawa (✉)
Food and Agriculture Organization of the United Nations (FAO), Rome, Italy

H. Kanamaru
Food and Agriculture Organization of the United Nations (FAO), Regional Office for Asia and the Pacific, Bangkok, Thailand

T. Iizumi et al. (eds.), *Adaptation to Climate Change in Agriculture*,
https://doi.org/10.1007/978-981-13-9235-1_11

The number of extreme climate-related disasters, including extreme heat, droughts, floods, and storms, has doubled since the early 1990s, with an average of 213 of these events occurring every year during the period of 1990–2016 (FAO 2018). These events harm agricultural productivity contributing to shortfalls in food availability, with knock-on effects causing food price hikes and income losses that reduce people's access to food. In order to minimize the impacts of climate change and extremes on agriculture and ensure food security especially in developing countries, the importance of appropriate adaptation has been recognized and emphasized in the global agenda of climate change.

Many researches have revealed the impacts of climate change on agriculture across the world and there is increasing evidence that adaptation actions in agriculture are occurring (IPCC 2014). In order to upscale adaptation to wider region such as national and regional scale, policy support is essential and there should be sufficient communication between scientists and policy makers to reflect the scientific findings into policy-making process and enhance the evidence-based adaptation planning. This paper introduces FAO's country support in agricultural adaptation by promoting dialogues between science and policy.

11.2 Adaptation in Global Agenda

Recently, climate change and agriculture have been framed in several major global agendas. The Sustainable Development Goals (SDGs), adopted in 2015 at the United Nations General Assembly, target to achieve 17 global goals by 2030 toward a better and more sustainable future for the world. SDG-2 aims to achieve zero hunger, and one of the targets (SDG 2–4) is to ensure sustainable food production systems and implement resilient agricultural practices. SDG-13 is about taking urgent action to combat climate change and its impacts, and it aims at strengthening resilience and adaptive capacity to climate-related hazards and natural disasters in all countries. One of the targets (SDG 13–2) is to integrate climate change measures into national policies, strategies, and planning.

In the same year, 2015, the parties to the UN Framework Convention on Climate Change (UNFCCC) adopted the Paris Agreement, a new framework for an internationally coordinated effort to tackle climate change. It aims to strengthen the world's commitment to reducing greenhouse gas emissions to keep a global temperature increase below 2 °C above pre-industrial levels, with an effort to limit it further to 1.5 °C. It also aims to strengthen the ability of countries to adapt to climate change "in a manner that does not threaten food production." Here, safeguarding food security is regarded as the fundamental priority, and "the particular vulnerabilities of food production systems to the adverse impacts of climate change" is recognized. The countries submitted nationally determined contributions (NDCs) and defined their required actions to reduce national emissions and to adapt to the impacts of climate change. The Paris Agreement demands a revision of NDCs at least every 5 years representing progression beyond the last NDCs.

On the other hand, the National Adaptation Plan (NAP) processes, established in 2010 under the UNFCCC, are domestic planning processes, whose objectives are to facilitate the integration of climate change adaptation, in a coherent manner, into relevant new and existing policies, programs, and activities, in particular development planning processes and strategies, within all relevant sectors and at different levels, as appropriate.

11.3 Types of Risk Assessment and Adaptation in Agriculture

As a specialized agency of the United Nations, Food and Agriculture Organization (FAO) has been supporting countries with tackling climate change both for mitigation and adaptation in agricultural sectors. In FAO's analysis of NDCs, it was evident that strengthening resilience and enhancing carbon sinks are key priorities for agriculture. At the same time, although not required, most countries included an adaptation component in their NDC in addition to mitigation commitments addressing the importance of adaptation to achieve the food security.

Before planning adaptation to climate change, climate risks need to be assessed, and there are two types of assessment approaches: top-down and bottom-up (Wilby and Dessai 2010). Top-down methods involve first downscaling climate projections from GCMs (General Circulation Models) under several greenhouse gas emissions scenarios. The resulting local scenarios are then fed into impacts models, such as crop growth models, before finally invoking adaptation measures to maximize benefits. Bottom-up methods, on the other hand, focus on reducing vulnerability to past and present climate variability, typically in the wake of an extreme event or disaster. The term bottom-up is used because the analysis begins with the factors and conditions that enable successful coping with climate-related threats at the level of individuals, households, and communities.

Thinking of adaptation types, on the other hand, it is recognized that incremental adaptation will not be enough and transformational adaptation will be required in some agro-ecosystems (Jones and Thornton 2009; Kates et al. 2012). Incremental adaptation maintains the essence and integrity of a system or process at a given scale, while transformational adaptation changes the fundamental attributes of a system in anticipation of climate change and its impacts (IPCC 2014). In the agriculture sector, improvements of crops (e.g., drought- or flood-tolerant variety) or on-farm management practices (e.g., irrigation timing and quantity, fertilizers, adjusting cropping calendars, use of weather forecast, seasonal climate forecast, and agrometeorological advisories) can be considered as incremental adaptation. Transformational adaptation takes a variety of forms, such as switching crop types, shifting locations for producing certain crops and livestock, shifting farming systems new to an area, exploring alternative livelihood strategies, etc. (Rippke et al. 2016).

11.4 Enhancing the Evidence Base for Adaptation Planning

Planning adaptation in national policy, especially transformational adaptation, requires a robust climate evidence: climate risks and vulnerabilities of the agro-ecosystem have to be assessed, ideally using an integrated approach of top-down and bottom-up methods. The evidence at the local scale justifies the choice of adaptation options and investments at the project level. Evidence at a larger spatial scale, such as sub-national to national, form the basis for adaptation planning and policies at the national level. Climate change adaptation is a long-term iterative process. Countries need to have the capacity to complete the full process of adaptation, from producing evidence, planning policies, developing and implementing projects, and monitoring and evaluating the progress. Most of the evidence can be highly scientific but they need to be well interpreted and linked to the policy making.

We identify five main types of evidence that facilitate evidence-based adaptation planning. The first is the evidence from the past: historical climate trends, including extreme weather events, and their impacts on agriculture. The second is the future: projection of climate and its impacts on agriculture. The third is characterized risks and vulnerability to climate change and social and environmental factors. The fourth is identification and appraisal of potential adaptation practices. They include on-farm trials and desktop studies such as cost-benefit analysis and biophysical assessments. Lastly, implemented adaptation practices need to be monitored and evaluated to assess their effectiveness for the next iteration in the adaptation pathway.

There are several hurdles to overcome when implementing evidence-based adaptation in developing countries. First, many assessment studies focus either on small-scale area, such as field experiment, or larger scale of continent or global scale impacts, but they are not informative at national and sub-national scales. Second, many assessment tools exist, but they are developed as research tools to answer academic questions in their own discipline, and are not designed for use to answer policy questions. Furthermore, generally speaking, the capacities of developing countries and their experts in producing evidence are insufficient, particularly in least developed countries. Most of the research is conducted by international scientists with only minimal involvement of local researchers.

11.5 MOSAICC (Modelling System for Agricultural Impacts of Climate Change)

11.5.1 Objectives of the Modelling System

In an attempt to fill the gaps above, FAO has been developing a capacity development tool, Modelling System for Agricultural Impacts of Climate Change (MOSAICC) since 2008. One of the features which differentiates MOSAICC from other tools is that it is designed purely as a capacity development tool which can be easily run by the local experts in developing countries with minimum requirement of input data. MOSAICC employs the top-down method to assess climate risks. It is a web-based platform, which allows users to make climate projections for the future and assess their impacts on crop, hydrology, forestry, and economy sectors of a user's interests.

All components of MOSAICC run on a server and exchange data through a central geospatial database. This system design brings together very different models that are usually run independently by separate groups of researchers. MOSAICC facilitates and fosters collaboration of researchers from different disciplines who tend to work only in their own domains.

11.5.2 Design

MOSAICC consists of five main components of the models (Fig. 11.1):

- Climate component for statistical downscaling to produce climate projections from General Circulation Models (GCMs)
- Crop component to simulate crop growth and future crop yields
- Hydrology component to estimate river water resources
- Forestry component to simulate biomass and tree species distributions
- Economy component to assess the effect of changing yields and water availability on national economies

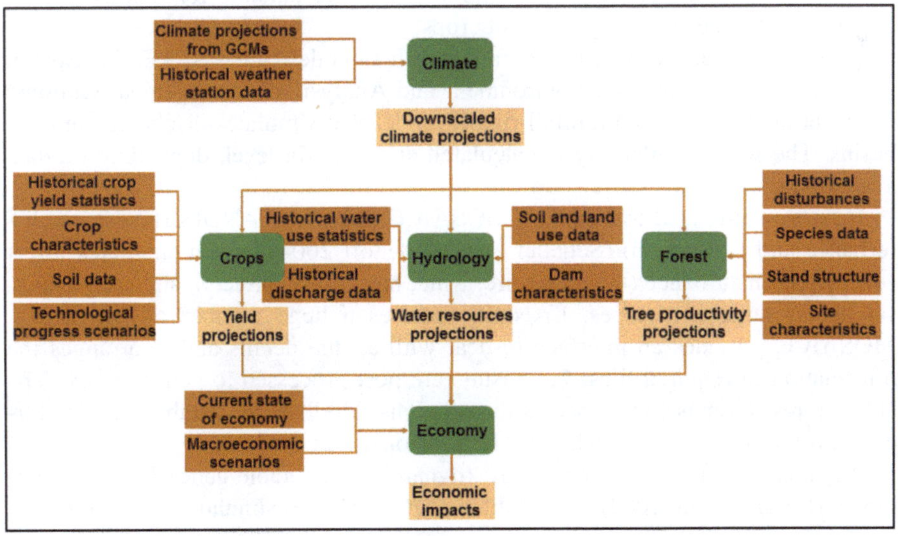

Fig. 11.1 MOSAICC components

Each component provides one or more models. There are also cross-component tools available, such as spatial interpolation, grid area analysis, and cell statistics. The models chosen to be integrated into MOSAICC are relatively simple and robust, and can run with input data of limited quality and availability in different ecosystems. Digital elevation model (DEM), land use, soil properties, weather data, hydrological data, crop yield statistics, and crop parameters are examples of data that are necessary for MOSAICC simulations. The participating models and tools are open source. As a result, MOSAICC is free and highly transferable to many different countries in diverse agroecological zones.

The locations for the weather stations and their time series data should be uploaded to perform statistical downscaling. The downscaling process is external to MOSAICC because it requires huge computation resources, but MOSAICC provides an interface to interact with it efficiently. There are several statistical methods available for climate downscaling (Gutiérrez et al. 2012).

The spatial interpolation operation addresses the problem of limited number and coverage of weather stations in developing countries. The optimized AURELHY (Analyse Utilisant le RELief pour les besoins de l'Hydrométéorologie) algorithm (Bénichou and Le Breton 1987) facilitates subsequent model simulations by gridding climate data as well as aggregating to administrative levels, as necessary.

There are several models and tools for assessing crop yield: the planting dekad model (PLD) (Franquin 1973), the water balance model (WABAL) (Frère and Popov 1986), and the crop water productivity model (AQUACROP) (Steduto et al. 2009). AQUACROP simulates crop yield and is usually used in specific locations because it requires a number of data collected in the field. WABAL is a simpler model with limited requirements of input data and produces crop-specific water balance variables as outputs. The variables are used to construct statistical models to simulate crop yield. The WABAL approach is more suitable for assessments at larger spatial scale and many different crops.

Hydrological assessment can be done with a model called STREAM (Spatial Tools for River Basins and Environment and Analysis of Management Options) (Aerts et al. 1999). It is a rainfall-runoff model that simulates discharges in river basins. The water availability is calculated at sub-basin level, depending on data availability.

For forestry-related assessment, MOSAICC provides LANDIS (landscape disturbance and succession) (Scheller and Mladenoff 2004), which simulates forest succession, disturbance (including fire, wind, harvesting, insects), and seed dispersal across large landscapes. LANDIS requires a huge number of parameters. MOSAICC provides an interface to deal with all the details and re-arranges the information in required files. The results are post-processed to generate key variables: forest biomass, tree species distributions, biodiversity, establishments, forestry evolution, Leaf Area Index (LAI), and non-wood products.

As an economic model, the DCGE (dynamic computable general equilibrium) model (Lofgren et al. 2002) is available in MOSAICC. It simulates the current and future economy under different climate projections. The model distinguishes the national economy and that of the rest of the world, between which goods and services are exchanged. The model uses crop yields and water availability generated by agronomists and hydrologists as shocks to the national economy. The main outputs are macro indicators (GDP), domestic market variables, external trade variables, and prices.

Key outputs of MOSAICC simulations are future projected values of these different variables simulated by each model. All the models can use the data generated from other models through a central geospatial database. The user works on MOSAICC with a web browser to connect to the MOSAICC server over the Internet. Data, models, and results are all on the server. Nothing is required on the user's computer. The systems installed in countries can be easily upgraded remotely by MOSAICC developers.

11.6 Implementation of MOSAICC in Countries

Up to date, MOSAICC has been successfully implemented in eight countries: Morocco, the Philippines, Peru, Indonesia, Paraguay, Malawi, Zambia, and Uruguay. In order to ensure that local researchers use MOSAICC to produce information that are useful for stakeholders, we emphasize a country-driven process for implementing MOSAICC. Typically, the implementation process takes the following four steps: needs and capacity assessment, formation of the working group, model simulation and analysis, and communication with decision-makers.

11.6.1 Information Needs and Capacity Assessment

A typical implementation of MOSAICC in a new country starts with a stocktaking exercise of existing information about climate change impacts on agriculture in the country. National ministries, such as ministry of agriculture and national meteorological agency, and relevant research institutions are consulted for this purpose. They are responsible for developing national climate change adaptation policies and programs, such as National Adaptation Plans, national climate change policies, National Communications to the UNFCCC, and agriculture investment plans, and they provide their views about what kind of information will be useful to produce. In parallel with information needs assessment, country's technical and institutional capacities to create the information required are assessed across ministries, national research institutes, and universities.

11.6.2 Interdisciplinary Technical Working Group and Designing the Study

If MOSAICC appears to address country's information and capacity gaps, we start forming an interdisciplinary technical working group that is composed typically of ministries, national research institutes, and universities. The subject experts in the group will be responsible for running simulations with each component of MOSAICC, and the group is supervised by the project steering committee in the country. They discuss and decide on the study objectives and design, such as time periods, targeted crops and the study areas of their interest, to ensure that the results address properly the policy questions and stakeholder needs. In this process, the group members have an opportunity to reflect on country's context, and to build a common understanding about what would constitute a successful adaptation to climate change in the agriculture sector, and what kind of information MOSAICC should produce in support of achieving the goal. This process is typically guided by policy makers in the working group.

Data collection is a time-consuming process and it is one of the largest challenges we face in many countries. MOSAICC requires relatively small amount of data as inputs to models, but data are often scattered across different offices, and not in a format suitable for computer processing. We also make sure that the data providers agree on sharing the data with all members of the technical working group so that a truly collaborative research is possible.

11.6.3 Simulations Through Capacity Development

Capacity development is another important focus in our MOSAICC implementation strategy. Adaptation planning should be periodically reviewed with new evidence, science, and outcomes from adaptation interventions. The built capacities of country experts to carry out science work that forms an evidence base about climate impacts and adaptation are key to a sustainable policy planning process. We provide extensive training programs to the identified local experts for use of each component of MOSAICC. At least 1 week of training per component is usually provided, and the trained experts are supported throughout the process to accomplish simulation studies and produce a policy recommendation.

The idea is that country experts can perform simulations using their country's own data in support of national planning. Once they built their national capacity, they can re-do the similar analysis whenever required. It takes about 3 months (per component) for experts to perform simulations provided dedicated researchers are assigned to the task.

11.6.4 Communication of Results

Running simulations is only a part of climate impact studies. The simulation results need to be analyzed, interpreted, and visualized for stakeholders. They would inform policy makers of which areas/subsectors/crops/basins/forest species are more vulnerable than others. This evidence-based information would support adaptation planning and allow strategic resource allocations, investment programs, research and development, and prioritization of adaptation interventions.

One of the tasks of the technical working group is to ensure the clear communication of implications of model outputs to aid policy processes. Communication of the results can take a number of other forms: presentation in conferences, paper and electronic publications, and website. MOSAICC is designed to publish results from the simulation server in a seamless manner as graphs/maps to the web server.

In order to reflect the analysis to the policy, there should be further discussion between the government and stakeholders to decide on the most appropriate adaptation practices which will be supported and approved by all the stakeholders. It is important that the governments identify and tackle the existing barriers, if there are any, which limit the adoption of the adaptation practices before designing and implementing new policy measures (Wreford et al. 2017).

11.7 Examples of Country Implementation

In this section, we introduce three case studies. The Paraguay (2015–2018) case focuses on scientific results while the Malawi (2015–2018) case describes the process of country-driven implementation. Further the Philippines (2012–2015) and Peru (2012–2015) cases focus on the establishment of the working group and the scheme of South-South cooperation.

11.7.1 Paraguay

Climate change impact assessment in the agriculture sector in Paraguay was conducted as part of the Analysis and Mapping of Impacts under Climate Change for Adaptation and Food Security (AMICAF) project phase 2 (2015–2018), through South-South cooperation (SSC). Climate downscaling was conducted based on the historical meteorological information from 12 meteorological stations for the entire country. The historical reference time considered was 1981–2010. The models showed that a reduction in precipitation is expected for the time period from 2041 to 2070, while temperatures (maximum and minimum) are expected to increase. The models showed a range of possible decrease in precipitation from 2.40% to 10.24% under RCP4.5 while a possible decrease in precipitation for RCP8.5 was

3.27–15.92%. RCP stands for Representative Concentration Pathway of atmospheric greenhouse gas (GHG) and RCP8.5 means higher CO_2 concentration scenario than RCP4.5. Projected temperature increases are from 1.8 to 2.7 °C (Tmax) and 1.8 to 3.3 °C (Tmin) under RCP4.5 and from 2.3 to 3.3 °C (Tmax) from 2.2 to 4.0 °C (Tmin) under RCP8.5.

The impact assessment of climate change on crops included the analysis of historical yields and projected trends for the future from 2041 to 2070. Eight crops were selected for this analysis: sugarcane, common beans, cassava, corn, wheat, soybean, irrigated rice, and rainfed rice. The results showed great heterogeneity in terms of future impacts on the yield of crops at the department level due to climate change, with both increasing and decreasing projections for different departments and crops. For several crops, no significant differences between historical and expected future yields were reported, or different GCM projections lead to inconsistent results. Figure 11.2 shows the result for cassava ("mandioca"). This crop, strongly associated with family small-scale agriculture, could significantly increase its yield in 5 departments out of 17, and the yield increase was consistent in both RCPs (4.5 and 8.5).

In summary, negative impacts of climate change are expected for sugarcane in five departments, soybeans in three departments, and upland rice in two departments. Interestingly, cassava shows a positive impact of climate change on yields for approximately the same region for which negative impact is observed for sugarcane. Common beans yields are expected to increase in three departments, and decrease in three other departments. We can conclude that higher climate change risks concern sugarcane, soybeans, and upland rice, while wheat, maize, and common beans show no significant changes or both positive and negative changes. We can consider these crops as the most resilient, while cassava yields appear to mainly benefit from climate change effects. These Paraguay results are a good example of evidence that facilitates transformational adaptation planning with potential options such as shifting cultivating areas of key crops within a country.

11.7.2 Malawi

Located in sub-Sahara Africa, agriculture in Malawi has been affected by climate change. In the 2015/2016 rainfall season, for example, Malawi was hit by a prolonged dry spell, and agriculture was severely affected by the drought. It was estimated that this season saw 12.4% decline from the 2014 to 2015 season in overall food production, which was already down by about 30% (due to 2015 floods) compared to the 2013/2014 season (Government of Malawi 2016).

The Government of Malawi is committed to taking action to tackle food security and climate change challenges. The NAP process was launched in 2014 to provide medium- to long-term options for Malawi to address adaptation needs. In 2017, Malawi Growth and Development Strategy (MGDS) III was established to move the country to a sustainable development growth path. MDGS III recognizes that climate change has adverse impacts on the agriculture sector.

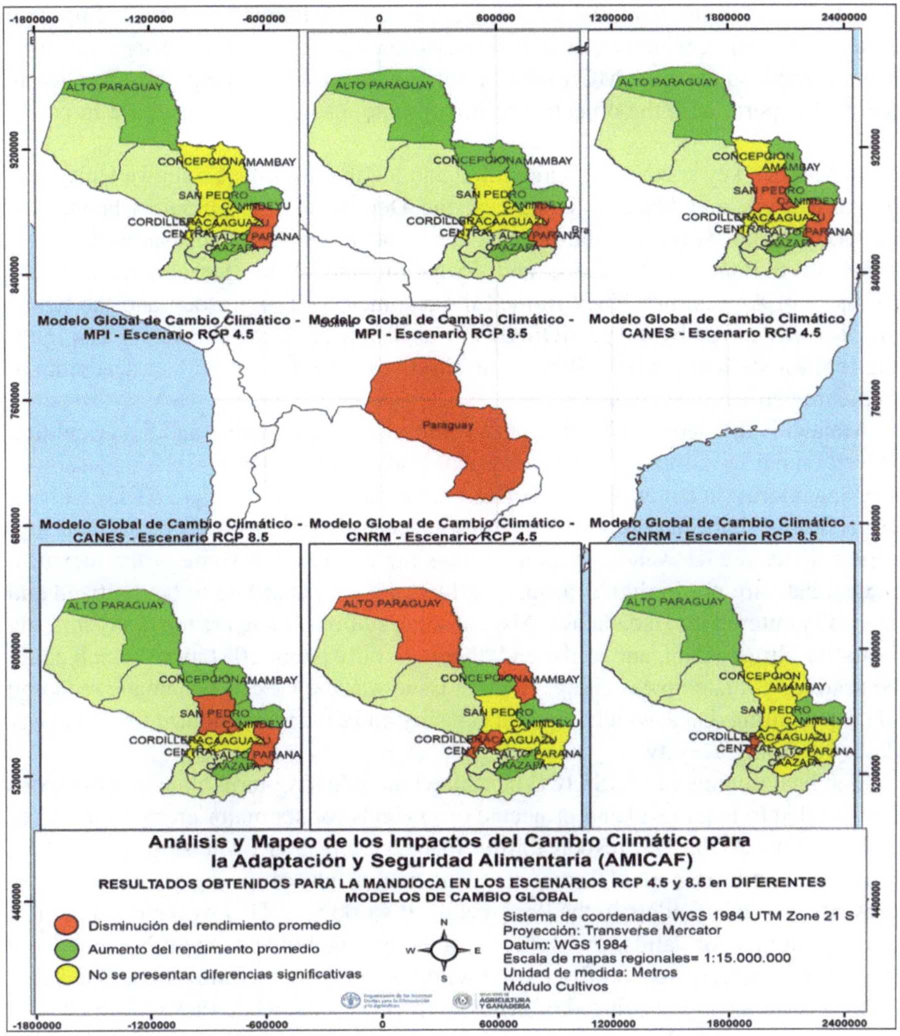

Fig. 11.2 Projected yield changes of cassava for the future (2041–2070) compared to the historical reference time period (1981–2010) under three GCMs (MPI, CANESM, and CNRM) with two RCPs (4.5 and 8.5) in Paraguay. Red area shows significant decrease in yield, green area shows significant increase, and yellow area shows no significant changes. If the change in average crop yield for the future is more than one standard deviation of that of the historical reference period, this change was defined as significant

In 2015, FAO began supporting Malawi to integrate the agriculture sectors into the NAP process. FAO is also a member of the NAP Core Team – the formal coordinating mechanism at national level, driving this process and reporting to the National Climate Change Technical Committee (NCCTP). A stocktaking exercise found that decisions at ministry level were reached without being backed up with data and evidence, which made it difficult to monitor policies for success or failure. FAO started supporting MOSAICC implementation to develop the capacity in national experts with the objective to incorporate a strong evidence base in policy making.

In Malawi, key stakeholders identified and involved are the Lilongwe University of Agriculture and Natural Resources, the Department of Climate Change and Meteorological Services, the Departments of Agriculture Research Services, Agriculture Planning Services, and Irrigation and Water Development in the Ministry of Agriculture. They formed the technical working group and the whole process was driven by them. MOSAICC was also used in the University as a training tool for students, who will be contributing to the relevant work in academia or in the ministries in the future.

Malawi is divided into 28 districts as administrative division, and 8 Agricultural Development Divisions (ADDs) for agricultural purposes. The stakeholders showed a strong interest in conducting crop assessments based on these eight ADDs with six crops: maize (hybrid, composite, local), sorghum, groundnuts, rice, beans, and soybeans. There are 41 weather stations across the country, but some of the meteorological data are not in digital format, and records on paper had to be digitized and manually entered into a database. Also, at some stations, there are too many missing years for climate data, and in the end, there are only about 20 stations which could be used for climate downscaling. Another issue was the lack of systematic recoding of crop planting dates, which required capacity and time to harmonize for use in any data processing activity.

The final outputs of MOSAICC in Malawi include long-term climate projections downscaled to local level and projected crop yields for six major crops across eight ADDs up to 2070. The responsibility, management, and ownership of the data, tools, and results remain in the country. The final results of impact assessment were presented to stakeholders in the final technical workshop. This workshop served as the opportunity for validation of these results by stakeholders, and the results were treated as a material to start discussion within and across the ministries. The stakeholders were mostly convinced of the assessment results, while they disagreed with some of the outputs. It was pointed out that several key management practices such as irrigation were not well considered, which resulted in misleading projections of crop productions in the future. The Department of Agriculture Irrigation and Water Development particularly showed a great interest in the simulation outputs, as the results may directly affect their plan for selecting the locations of new irrigation systems in the very near future. An examination of an alternative crop to maize for the future was another topic of interest. The nation-wide analysis with sub-national (ADD) disaggregation gave them an indication about which crops may be supported in which areas in the country. Needless to say, it was pointed out that the crop preference of people and market should be taken into consideration.

Currently, the Malawian team is in the process to finalize the technical report, reflecting the discussion at the final technical workshop, and disseminate their results to relevant stakeholders including several ministries. While there are various degrees of agreements in the ensemble of simulation results, they were able to identify consistency in climate projections in specific areas, crops particularly sensitive (or insensitive) to these changes, or areas particularly impacted for most of the crops, for instance. The nation-wide information, directly related to policy-relevant administrative boundaries, provides a new basis for improving adaptation measures (e.g., new crops of relevance, irrigation options) for the government.

Several important lessons were learned in the implementation process. Participating experts referred to the lack of human resources within the ministries as a limiting factor in maintaining momentum to complete the MOSAICC process. Strong encouragement of the team leader and senior management of the ministries to push the process forward is necessary to successfully coordinate the various components and to keep the experts engaged throughout. The final objectives and key milestones should be clearly laid out to incentivize the work. The utility of learning these models and the transferability of the skills gained should be clear to the experts as additional motivations to be involved in MOSAICC exercises.

11.7.3 The Philippines and Peru

Climate change impact assessment in the agriculture sector in the Philippines and Peru was conducted as part of the AMICAF project phase 1 (2012–2015). The phase 2 (2015–2018) targeted Indonesia and Paraguay using the scheme of South-South cooperation (SSC) where the Philippines and Peru also supported the implementation in the other two countries.

In Peru, the Ministry of Agriculture and Irrigation (MINAGRI) was identified as the main stakeholder and its Vice-minister chaired the steering committee of the project that implemented MOSAICC. For the technical working group, the National Meteorological and Hydrological Services, the National Agrarian University in La Molina, and the Office of Economic and Statistical Studies in the Ministry of Agriculture, were found to possess relevant knowledge and skills to do the analysis in each component. Peru was interested in extending the projections up to the end of the twenty-first century, with a set of 29 crops at the region level, and 16 river basins that represent different agroecological systems.

In the Philippines, the Department of Agriculture, Philippine Atmospheric, Geophysical and Astronomical Services Administration, Philippine Rice Research Institute, and National Institute of Geological Sciences in the University of the Philippines were the main members of the technical working group. The Philippines decided to examine climate change impacts up to mid-twenty-first century. They had already conducted dynamical downscaling, and hence, the climate statistical downscaling work is considered to complement it and to provide up-to-date information with a new set of climate projections (CMIP5). Their MOSAICC crop and hydrology work were designed to highlight differentiated impacts of climate change in different locations across the whole country with a focus on rice and corn at the province level, and 24 river basins.

The work in the Philippines was presented in a national project conference hosted by the Department of Agriculture, with wide participation from other Departments, Climate Change Commission, research institutes, universities, international development agencies, NGOs, and media. The nation-wide assessment work was highly appreciated and form a basis for NAPs in the agriculture sector, Philippine Development Plan, National Climate Change Action Plan, and other policy processes.

As part of SSC in AMICAF phase 2, Philippines experts trained in phase 1 travelled to Indonesia in 2017 for the technical workshops where preliminary results of MOSAICC simulations were presented. They shared their experiences in interpreting the results, summarizing them in policy context, and communicating them to stakeholders at the community level as well as at the national level. Peruvian experts were engaged from an earlier stage of MOSAICC implementation in Paraguay. They presented both scientific aspects and the policy context of the Peruvian MOSAICC study, and also discussed formulation of the interdisciplinary working group, inter-ministerial coordination, and other topics regarding project management, which informed the design of the Paraguay MOSAICC exercises.

11.8 Conclusions

Agriculture is widely recognized as one of the most vulnerable sectors to climate change, and it requires urgent action, from the farm to global level. Global agendas that frame the challenges are in place, especially to support developing countries. In view of promoting transformational adaptation, there is need to enhance the evidence base for adaptation planning, climate risks, and vulnerability in agriculture at the right spatial and temporal scales and to further develop the capacity in the country to produce the relevant information and communicate with policies.

FAO has developed MOSAICC and transferred it to countries to address the gaps. It typically employs the method of top-down risk assessment at national level, and its simple nature contributes to building more sustainable institutional capacities within countries, hence improving ownership of the process of assessment. The locally developed evidence that is relevant to national context supports policy discussions at the national level. The assessment, conducted by the national experts, serve as a basis and a trigger for inter-ministerial discussions.

The involvement of the governmental people brings another benefit: promotion of evidence-based adaptation within the community of practice. They attend regional and international workshops on climate change adaptation frequently, and exchange information with other countries. A successful experience of a country in implementing agricultural policies based on robust evidence motivates other countries to do so in their own country too. South-South cooperation has also been facilitated by FAO between Philippines and Indonesia and Peru and Paraguay, where the lessons learned in one country are communicated to the other, resulting in successful implementation.

There are high expectations for science community to contribute to the climate change and agriculture agenda by conveying scientific information to national policy making. To do so, application-oriented research and tools have to be designed and whose outputs should be translated into clear messages for decision-making in national and sub-national policies. A neutral forum where scientists and policy makers in adaptation community can exchange information for advancing climate change risks and vulnerability assessment will be highly useful. For this purpose, FAO is currently developing a new version of MOSAICC only with public data and simplified methodologies, for quick climate impact and vulnerability assessments by national and international experts with minimum training. The current implementation of MOSAICC requires extensive training and support from FAO and input data from countries, although minimum; hence, it doesn't allow us to support many countries at the same time. The new MOSAICC will enable us to respond to increasing requests from countries, and also to address the needs from adaptation investment projects, such as those funded by the Green Climate Fund, that require robust evidence about climate vulnerabilities for justification and prioritization of investments. Our work was made possible by close partnership with universities and research institutes in climate and agriculture sciences. There will be even more opportunities for further strengthening collaboration between the academia and the development community in the work of climate risks, vulnerability and adaptation assessment.

Acknowledgments The development of MOSAICC was supported initially by an EU/FAO Program. FAO projects, funded by Japan, Belgium, Netherlands, and Germany, provided financial support for MOSAICC implementation in countries. We would like to express our gratitude to all the technical partner institutions and individuals who developed the models that constitute MOSAICC: University of Mons (Belgium), Institute for Environmental Studies, Free University of Amsterdam, Water Insight (the Netherlands), and University of Cantabria (Spain). The authors also would like to acknowledge the work of their colleagues in national ministries, research institutes, and universities who carried out climate change impact studies in each country and FAO country offices and project staff who supported their work. We thank FAO colleagues in Rome who contributed to the development and implementation of MOSAICC.

References

Aerts JCJH, Kriek M, Schepel M (1999) STREAM, spatial tools for river basins and environment and analysis of management options: set up and requirements. Phys Chem Earth Part B Hydrol Oceans Atmos 24:591–595

Bénichou P, Le Breton OA (1987) AURELHY: Une méthode d'analyse utilisant le relief pour les besoins de l'hydrométéorologie. Journées Hydrologiques de l'ORSTOM à Montpellier 2:299–304

FAO, IFAD, UNICEF, WFP and WHO (2018) The state of food security and nutrition in the world 2018. Building climate resilience for food security and nutrition. FAO, Rome

Franquin P (1973) Analyse agroclimatique en régions tropicales. Méthode des intersections et période fréquentielle de végétation. Agronomie tropicale 28:665–682

Frère M, Popov GF (1986) Early Agrometeorological crop yield forecasting. FAO Plant Production and Protection, FAO, Paper no 73, p 150. FAO, Rome

Government of Malawi (2016) Malawi drought 2015–2016 post-disaster needs assessment (PDNA)

Gutiérrez JM, San-Martín D, Cofiño AS, Herrera S, Manzanas R, Frías MD (2012) User guide of the ENSEMBLES downscaling portal. Version 3. Technical note 2/2012, Santander Meteorology Group, Spain

IPCC (2014) Climate change 2014: impacts, adaptation, and vulnerability. Part A: global and sectoral aspects. contribution of working group II to the fifth assessment report of the intergovernmental panel on climate change Field CB, Barros VR, Dokken DJ, Mach KJ, Mastrandrea MD, Bilir TE, Chatterjee M, Ebi KL, Estrada YO, Genova RC, Girma B, Kissel ES, Levy AN, MacCracken S, Mastrandrea PR, White LL (eds), pp 485–533. Cambridge University Press, Cambridge/New York

Jones PG, Thornton PK (2009) Croppers to livestock keepers: livelihood transitions to 2050 in Africa due to climate change. Environ Sci Policy 12:427–437

Kates RW, Travis WR, Wilbanks TJ (2012) Transformational adaptation when incremental adaptation to climate change are insufficient. Proc Natl Acad Sci USA 109:7156–7161

Lofgren H, Thomas M, El-Said M (2002) A standard Computable General Equilibrium (CGE) model in GAMS, a standard Computable General Equilibrium (CGE) model in GAMS, International Food Policy Research Institute 5, International Food Policy Research Institute

Rippke U, Ramirez-Villegas J, Jarvis A, Vermeulen S, Parker L, Mer F, Diekkruger B, Challinor AJ, Howden M (2016) Timescales of transformational climate change adaptation in sub-Saharan African agriculture. Nat Climate Change 6. https://doi.org/10.10387/NCLIMATE2947

Scheller RM, Mladenoff DJ (2004) A forest growth and biomass module for a landscape simulation model, LANDIS: design, validation, and application. Ecol Model 180:211–229

Steduto P et al (2009) AquaCrop – the FAO crop model to simulate yield response to water: I. Concepts and underlying principles. Agron J 101(3):426–437

Wilby RL, Dessai S (2010) Robust adaptation to climate change. Weather 65(7):180–185

Wreford A, Ignaciuk A, Gruère G (2017) Overcoming barriers to the adoption of climate-friendly practices in agriculture. OECD food, agriculture and fisheries papers, no. 101. OECD Publishing, Paris. https://doi.org/10.1787/97767de8-en

Chapter 12
Use of Seasonal Climate Forecasts in Agricultural Decision-Making for Crop Disease Management

Kwang-Hyung Kim, Yonghee Shin, Seongkyu Lee, and Daeun Jeong

Abstract Recently, seasonal climate forecasts (SCFs) and their advances have gained increasing attention in agricultural communities, specifically because of their potential to improve climate risk management by increasing preparedness and thus enhance agricultural and economic outcomes. Seasonal predictions of crop diseases and insect pests that provide timely and accurate forecasts are especially valuable not only to farmers and extension workers (i.e., to inform their crop management decisions) but also to governments (i.e., to increase national-level disaster preparedness). In this study, we used a case study in Bicol, Philippines, to introduce an array of implementation strategies to facilitate the use of SCFs in the agricultural sector. To demonstrate the full potential of SCFs in the Bicol region, we developed and applied seasonal disease predictions for rice, with sequential activities that included a baseline study on disease epidemics in the target area, the examination of available SCFs, the development of a decision support system for seasonal disease predictions, and an evaluation of this system using SCF hindcasts. Finally, we disseminated the resulting seasonal disease predictions in the target area using an agromet bulletin. The present study demonstrated a successful example of a developmental framework for the application of SCFs to agricultural decision-making with the support of relevant SCF-linked agricultural models. These implementation strategies, in combination with the lessons learned, can help guide prospective efforts of establishing similar climate services that utilize SCFs in developing countries to improve the outcomes and thus lead to enhanced and sustainable food security.

12.1 Introduction

Extreme weather and other climate change-related events can lead to loss of life, injury, or other health issues, as well as to the damage and loss of property, infrastructure, livelihoods, service provisioning, and environmental resources, and thus

K.-H. Kim (✉) · Y. Shin · S. Lee · D. Jeong
Department of Climate Services and Research, APEC Climate Center, Busan, Korea
e-mail: kh.kim@apcc21.org

T. Iizumi et al. (eds.), *Adaptation to Climate Change in Agriculture*,
https://doi.org/10.1007/978-981-13-9235-1_12

impact all sectors that are pertinent to sustainable development (IPCC 2012, 2014). Among these sectors, agriculture is possibly the most sensitive and vulnerable sector to climate change and extreme weather variations. Since food security depends on agricultural systems that are resilient against climate-driven production failures, there is an urgent need to identify climate service-based adaptation strategies in agricultural systems (FAO 2015). Ideally, climate services transform climate information into practical applications by helping decision-makers make more informed and effective decisions. However, for such outcomes to be realized, climate services not only need to consider the reliability and accuracy of climate information and its relevance to adaptive decision-making, but also, using a holistic approach, ensure strong stakeholder engagement and ownership within the climate services themselves (Challinor 2009).

Reporting on and utilizing climate variability in climate services requires a sound understanding of the drivers of the variability and the extent to which these can be predicted. Improvements in our understanding of the interactions between the atmosphere and sea or land surfaces, advances in modeling global climate, and substantial investment in monitoring and observations now provide some degree of predictability for climate fluctuations, even months in advance, in many parts of the world (Goddard et al. 2001). In recent years, dynamic *g*eneral *c*irculation *m*odels (GCMs) have become mainstream tools to deliver *s*easonal *c*limate *f*orecasts (SCFs), due to their promising predictive skills (Power et al. 2007; Syktus et al. 2003). In addition, considerable efforts have been made to develop methods to quantify their skill levels and maximize their usability for potential end-users in the agricultural sector.

Due to the inherent uncertainty in climate prediction, SCFs are reported in probabilistic tercile terms (i.e., below-, near-, and above-normal categories). Understanding the nature of probabilistic forecasts is critical in utilizing climate information in various applications across sectors, including the agricultural sector. Unlike weather forecasts, SCFs can be predicted with a longer lead time for an upcoming season (i.e., 3–6 months). A longer forecast lead time not only informs short-term decisions, such as sequential seeding times, harvesting for short-term storage, or irrigation according to anticipated water stress, but also long-term decisions, such as the management of water resources and infrastructure, determination of labor recruitment, selection of crops and varieties, and determination of optimal cultivation areas and cropping systems. However, it is very difficult to provide useful information to inform farm-level decisions and policy-level interventions with the exclusive use of SCFs. For example, decision-makers are interested in the consequences of seasonal climate variability, rather than in the climate variability itself. Therefore, it is important that forecasts be appropriately "contextualized" before they can positively influence decision-making (Meinke and Stone 2005). One such way to contextualize forecasts is to link them with impact models, e.g., agriculturally relevant information (such as crop yields), to strengthen the potential impacts of the forecasts when presenting them to decision-makers. In fact, the emerging ability to probabilistically forecast upcoming seasons, in terms of both climate and its influence on agricultural systems, has started to influence decision-making to maximize benefits and minimize any potential climate-related risks during the growing season.

Technical and institutional obstacles often impede the integration of SCFs and agricultural models. For example, Hansen and Indeje (2004) argued that SCFs would only contribute to the improvement of agricultural risk management, if the spatial and temporal scale discrepancies between the GCMs and agricultural models are addressed. To date, many methods and systems, used to link SCFs with agricultural models, have been attempted and developed through interdisciplinary research (Capa-Morocho et al. 2016; Han et al. 2018; Martins et al. 2018). Attempts have also been made to produce reliable probability estimates of crop yields using multiple GCMs or ensemble predictions based on multiple crop models, or both, rather than single GCMs and crop models (Canal et al. 2017; Cantelaube and Terres 2005; Challinor et al. 2005; Iizumi et al. 2018; Rosenzweig et al. 2013). Participatory, cross-disciplinary approaches have been promoted to unite institutions, disciplines, and professionals to increase the uptake and reap the maximum benefits of agro-climate products (Hansen et al. 2011; Meinke and Stone 2005).

In this study, we introduced an array of implementation strategies for the use of SCFs in the agricultural sector using a case study in Bicol, Philippines. In the Philippines, SCFs can be predicted with high accuracy, especially for precipitation during the dry farming season, because they take the influence of El Niño-Southern Oscillation (ENSO) into consideration in many of the regions (Lyon et al. 2006; Mason and Goddard 2001). In addition, a strong relationship exists between the amount of seasonal precipitation and rice production in the Philippines (Koide et al. 2013). Therefore, SCFs with higher accuracy (skill) and enough lead time will presumably benefit farmers or extension workers, allowing for more opportunities to reduce climate risks, and thus improve regional (and possibly global) food security.

To demonstrate the potential benefits of SCFs in the Bicol region, we developed and applied seasonal disease predictions for rice. Our sequential activities included: (1) A baseline study on the disease epidemics in the target area; (2) Examination of available SCFs; (3) Development of a decision support system for the seasonal disease predictions; (4) Evaluation of the decision support system using an SCF hindcast (i.e., a retrospective forecast); and (5) Dissemination of the seasonal disease prediction in a target area. The aim of this study was to present a successful example of a developmental framework for the application of SCFs to all stages of agricultural decision-making; especially to demonstrate how decisions can be supported by linking SCFs (the inputs) to relevant agricultural models to generate agricultural consequences (the outputs). Furthermore, we have offered recommendations for SCF-based decision-making in the agricultural sector, to help ensure that prospective efforts to establish similar SCF-based climate services in developing countries result in better outcomes which lead to enhanced and sustainable food security.

12.2 Implementation Strategies: A Case Study

12.2.1 Baseline Study on Disease Epidemics in the Target Area

The Bicol region in the Philippines is characterized by extensive, rich flat land and agriculture forms the largest component of the economy (CountryStat Philippines 2013). The region is highly vulnerable to climate changes and extreme weather events, e.g., droughts and floods, generated by frequent typhoons and the seasonal effects of ENSO. This vulnerability translates to losses in agricultural productivity, which endanger the local economy as well as the regional food security. In 2014, the region suffered an extended drought and four typhoons, which resulted in a production loss of more than USD 53 million (Philippine Department of Agriculture Regional Office No. 5 2014). Additionally, the current impacts of ENSO are expected to continue across the Pacific and particularly impact food security in the region. The impacts of long-term climate changes pose serious uncertainties in the agricultural sector of this region; therefore, "climate services for agriculture" are urgently needed to improve local capacity to anticipate and respond to interannual and seasonal climate risks to ensure long-term food security.

The present study focused on rice-based agricultural systems in the Bicol region. An analysis by Hossain et al. (1995) showed that rice yield losses, due to agricultural disasters in the Philippines, amounted 945 kg/ha during the wet season (WS) and 1298 kg/ha during the dry season (DS). Apart from natural calamities, such as typhoons and drought, the highest contributors to the total yield loss were insects and diseases (26% in the WS and 16% in the DS); i.e., mainly tungro virus disease, stem borers, brown planthoppers, sheath blight, bacterial leaf blight, and rice bugs (Hossain et al. 1995). The prevention of these losses would, therefore, translate into increased rice yield.

Many of the locally reported rice diseases have been thoroughly investigated to prioritize target diseases. Disease prioritization is dependent on endemicity over the preceding years, as well as the economic and social impacts in the region. We clarified the disease epidemics and their economic and social impacts on farmers in the Bicol region by conducting a thorough literature review and by interviewing local partners. We identified the major rice diseases in the region and their control management practices by conducting multiple interviews with local partners from the Department of Agriculture - Regional Field Office No. 5 (DA-RFO5), municipal governments, and local state universities. Key questions for the interviews included: "What are the most socio-economically important diseases in your area?" and "Which diseases could be managed using the available control measures if their disease incidence was predicted prior to the start of the season?"

The Regional Center for Plant Protection under the DA-RFO5 provided long-term (more than 10 years) disease surveillance data that was collected at the provincial and municipal level. Local state universities also provided various project-related,

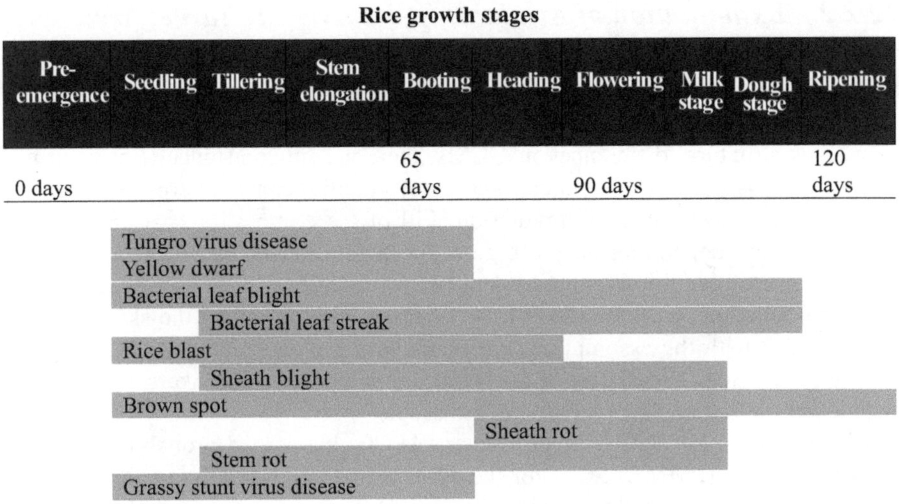

Fig. 12.1 Major diseases reported in the Bicol region and their period of occurrence within the rice cropping season

field-scale disease incidence data for model development. Interviews with farmers revealed the diseases that were most problematic to their rice farming and the available countermeasures. The major rice diseases in Bicol are summarized in Fig. 12.1, which includes their period of occurrence within the rice cropping season. Among the rice diseases reported in Bicol, rice tungro disease was perceived as the most important for rice crops; and bacterial leaf blight, sheath blight, and blast were ranked the next most important diseases. Other diseases were also mentioned; however, their impacts on rice yield were not considered to be significant.

A few target diseases were selected for use in the modeling. The selection of the final disease list was based on the baseline study and on the criteria of being not only economically important but also climatically sensitive; i.e., enable prediction by SCFs. This final disease list included: rice blast caused by the fungal pathogen *Magnaporthe oryzae*, sheath blight caused by the fungal pathogen *Rhizoctonia solani*, bacterial leaf blight caused by the bacterial pathogen *Xanthomonas oryzae*, and brown spot caused by the fungal pathogen *Bipolaris oryzae* (for more details, refer to The Rice Knowledge Bank, www.knowledgebank.irri.org). Disease outbreaks largely influenced by human activities were omitted from the selection process. In addition, since the modeling work requires tremendous time and effort, it was essential to determine the target diseases based on the availability and applicability of existing disease epidemiological models. As we couldn't find proper epidemiological models for rice tungro disease that are capable of simulating insect vector population dynamics driven by weather information, we excluded the tungro disease in the study.

12.2.2 Examination of Available SCFs Over the Target Area

SCFs were used as the input variables in disease epidemiological models to produce seasonal disease predictions. Due to fundamental variations in the initial conditions, as well as structural differences in GCMs, there are inherent uncertainties in the climate forecasts produced by them. These uncertainties can be interpreted either as an uncertainty level or as the prediction skill of the climate forecast of a certain region. In fact, depending on the region, the prediction skills of SCFs can vary widely. Several conditions must be met before an SCF can be translated into an improved outcome, e.g., the skill of the forecasts. We examined the skill over the target region with the assumption that SCFs with high prediction abilities would increase the potential usability of the information by decision-makers, while SCFs with lower prediction skills would be less usable.

Since 2007, the *APEC Climate Center* (APCC) has issued probabilistic multi-model ensemble (PMME) SCFs for the upcoming 3- to 6-month overlapping seasons (www.apcc21.org). These operational PMME forecasts by the APCC are an SCF in the form of tercile (the ordered distributions divided into three parts, each containing a third of the population)-based categorical probabilities, i.e., the probabilities of below-normal, near-normal, and above-normal categories (e.g., Boer 2005; Kharin and Zwiers 2003). In the present study, we examined the skill of the APCC PMME SCFs covering the Bicol region.

Monthly and seasonal PMME hindcast data from the APCC were used in the analysis. The seasonal PMME hindcast was comprised of rolling 3-month seasons and thus included all calendar months (i.e., JFM, FMA, MAM, AMJ, MJJ, JJA, JAS, ASO, SON, OND, NDJ, DJF). Both the monthly and seasonal hindcast data were provided in a 2.5 × 2.5° latitude/longitude grid format. The hindcast data from 1983 to 2005 were used. The lead time of the monthly data used in this study was up to a maximum of 3 months from the forecasting time point. The selection of GCMs for this composite was based on the availability of the longest and most continuous quality-controlled common hindcast datasets. An equal weighting approach was used to determine the contribution of each ensemble to the mean of the ensembles. Descriptions of the models used are presented in Table 12.1.

To verify the precipitation forecasts over the Bicol region, two datasets were obtained from (1) the Climate Anomaly Monitoring System and Outgoing Longwave Radiation Precipitation Index (CAMS OPI; Janowiak and Xie 1999) and (2) the Pili weather station, the synoptic weather station of the Philippine Atmospheric Geophysical and Astronomical Services Administration (PAGASA) in the Bicol region. The CAMS OPI dataset was comprised of precipitation estimates, for which observations from rain gauges were merged with estimates generated from a satellite product. The horizontal resolution of the CAMS OPI dataset was the same as that of the APCC PMME dataset. Weather observation data from 1979 to 2015 were obtained from the Pili weather station (13.555°N, 123.270°E), which is located within the 12.5°N, 122.5°E APCC MME forecast grid box (Fig. 12.2a).

Table 12.1 A description of the individual models used in the APEC Climate Center (APCC) probabilistic multi-model ensemble hindcast, and their ensemble sizes

Model name	Institution	Model resolution	References
CWB	Central Weather Bureau (Taipei)	T42L18	Liou et al. (1997)
GloSea5	Korea Meteorological Administration (Korea)	0.83°× 0.56°	Kang et al. (2011)
HMC	Hydrometeorological Centre of Russia (Russia)	1.125 × 1.40625°	Trosnikov et al. (2005)
JMA	Japan Meteorological Agency (Japan)	T95L40	Takaya et al. (2010)
MSC_CANCM3	Meteorological Service of Canada (Canada)	1.41 × 0.94°	Scinocca et al. (2008)
MSC_CANCM4	Meteorological Service of Canada (Canada)	1.41 × 0.94°	Merryfield et al. (2013)
NASA	National Aeronautics and Space Administration (USA)	288 × 181	Molod et al. (2012)
NCEP	Climate Prediction Center, NCEP/NWS/NOAA (USA)	T62L64	Kanamitsu et al. (2002)
PNU	Pusan National University (Korea)	T42L18	Ahn and Kim (2013)
POAMA	Centre for Australian Weather and Climate Research/Bureau of Meteorology (Australia)	T47L17	Lim et al. (2012)

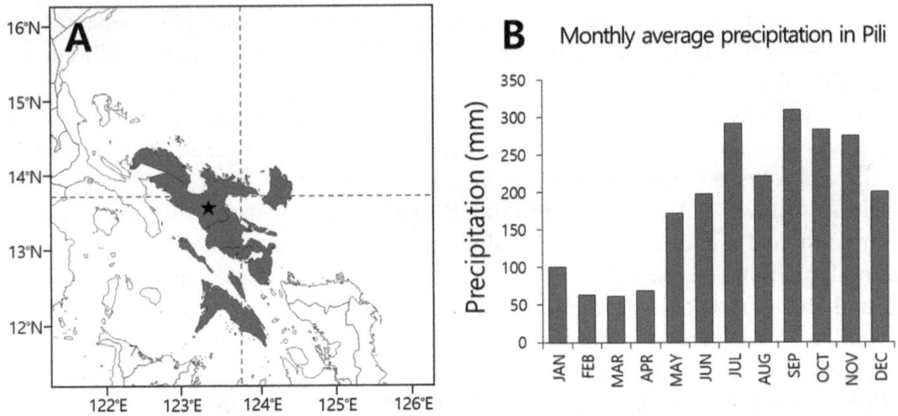

Fig. 12.2 (**a**) Location map and (**b**) monthly precipitation during 1979–2015 of the Pili weather station in the Bicol region, Philippines. (**a**) The black star, the gray regions, and the dotted grids on the map indicate the location of the Pili weather station, the Bicol region, and the APCC PMME data grid, respectively

The annual average precipitation in Pili, calculated from weather station data from 1979 to 2015, was 2246.9 mm. The monthly precipitation distribution is shown in Fig. 12.2b. Pili is characterized by a Type IV climate with relatively low precipitation from December to May and relatively high precipitation from June to

November (Koide et al. 2013). The APCC PMME data were in a 2.5 × 2.5° grid format, while the observed station data were obtained from the Pili weather station. Since it is less reasonable to directly compare the observed station data with the larger grid data, we used observation-comparable data in a grid format: the CAMS OPI. To test whether the CAMS OPI could be used to represent the Pili station data, we analyzed the correlation coefficients between the CAMS OPI and Pili station data. Although the correlation did vary among months and seasons, relatively high correlations (up to 0.79) were observed in most months and seasons, with a significance of 95%. Based on these results, we used the CAMS OPI for the verification of the APCC PMME SCFs, as a proxy for the observed precipitation data from the Pili station.

Ranked probability skill scores (RPSSs) were used to evaluate the prediction skill of the APCC PMME over Pili. Further details regarding the equations for RPSS can be found in the studies by Epstein (1969) and Murphy (1969, 1971). RPSS reached its maximum possible value at 1. The RPSSs for monthly PMME hindcasts with 1-month lead times were higher than those with 2- and 3-month lead times for all months except March, May, and June, and the RPSSs with 1- and 2-month lead times were all above zero (Fig. 12.3a). The average RPSS with a 1-month lead time was 0.172 and the scores for May, June, July, August, September, and November were below average. Similarly, the seasonal PMME hindcasts for the MJJ to OND seasons also had RPSSs below the average of 0.238 (Fig. 12.3b). The AMJ season was found to have the highest RPSS (0.426). The RPSS tended to decrease from April/AMJ until it fell below the average, and then increased again from September/SON. This annual cycle coincided with the WS (Jun to Nov) and DS (Dec to May) in Pili, respectively, indicating that the APCC PMME showed a higher skill during the DS in the region.

To further assess the skill of the APCC seasonal PMME hindcast, the Pili station data was directly compared with the PMME grid data for the period of 1983–2005. Skill was assessed based on a tercile hit rate analysis. Tercile hit rates were calculated as follows: (1) The 3-month rainfall totals from the station observation data were assigned a tercile (1, 2, or 3) value, depending on the percentage rank of the rainfall amount (i.e., less than 33.3%, tercile 1; between 33.3% and 66.6%, tercile 2; and greater than 66.6%, tercile 3) calculated over the entire period (1983–2005); (2) if the PMME tercile values matched the observed tercile values, then a "hit" was declared; and (3) the total number of hits was summed across the period of 1983–2005 to determine the "hit rate" for the APCC PMME hindcast.

The results of the tercile analysis are shown in Fig. 12.4. In the case of the tercile hit rates, a value equal to or less than 33.3% would indicate little or no skill. Values in excess of this threshold indicated potential skill, although they may not be statistically significant due to the relatively small sample size used. Except for the MJJ season, potential skill in predicting precipitation terciles was apparent in most seasons. The DS period from DJF to AMJ, especially, showed higher skill (up to 73.9% hit rate), partially because the impact of ENSO becomes more obvious during the DS than during the WS. Overall, the highest hit rate was 73.9% in the DJF season, which coincides with the peak of the DS rice planting in the Bicol region.

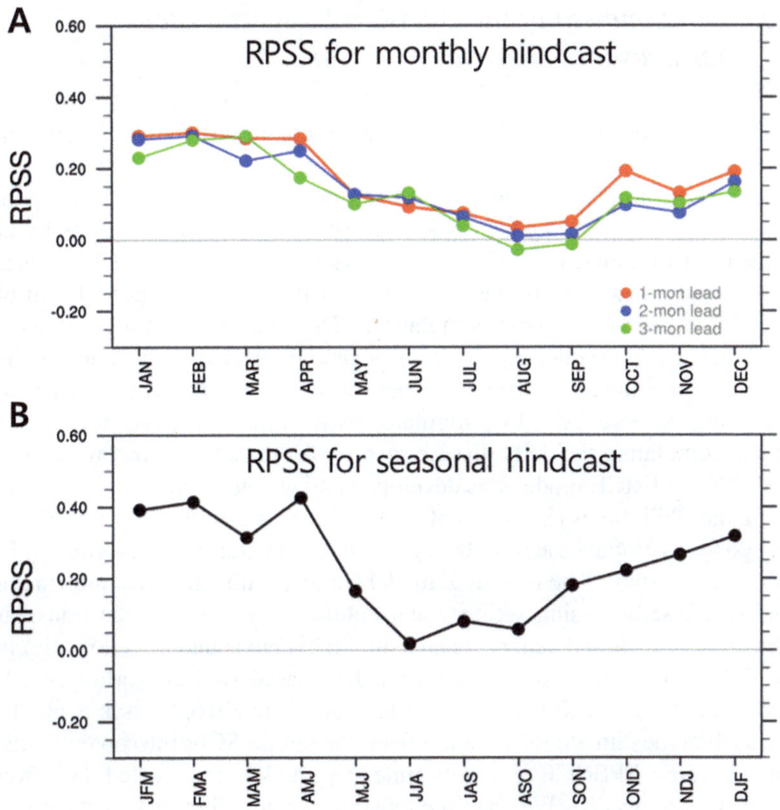

Fig. 12.3 Average ranked probability skill scores (RPSSs) for (**a**) monthly and (**b**) seasonal probabilistic multi-model ensemble hindcasts of precipitation over Pili during 1983–2005

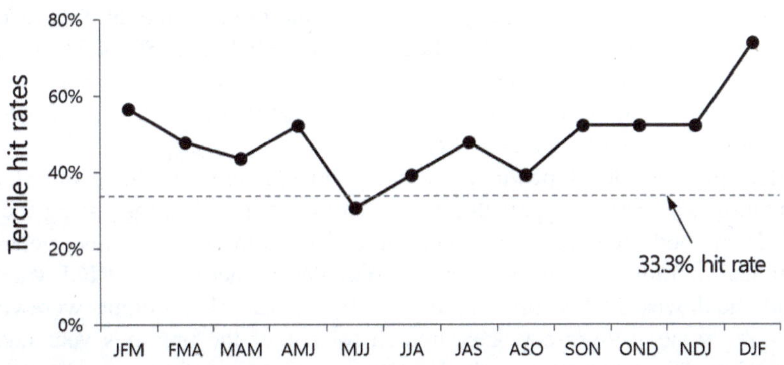

Fig. 12.4 Hit rates for the seasonal APEC Climate Center probabilistic multi-model ensemble (APCC PMME) hindcast of precipitation compared with observed weather data from the Pili weather station during 1983–2005. (Note: except for the MJJ season, most seasons of the APCC PMME hindcast showed more than 33.3% hit rates)

12.2.3 Development of the Decision Support System for Seasonal Disease Predictions

Crop diseases are primarily governed by weather conditions, including temperature, precipitation, wind speed, and humidity, with controlled or fixed management options. Disease forecast models have been developed to quantify the potential effects of weather on the intensity and occurrence of disease infections by pathogenic agents. To simulate potential disease risks for the four selected diseases in this study, we used an existing disease model, EPIRICE, for rice blast, sheath blight, bacterial blight, and brown spot simulations. Detailed infection algorithms of the EPIRICE related to weather input variables can be referred to the study by Savary et al. (2012). Potential disease infection risks were represented by the area under the disease progress curve (AUDPC) simulated by the model. The AUDPC was calculated by accumulating the daily disease severity for the entire growing season (i.e., % days). The EPIRICE model was developed and evaluated for the tropical region, including the Philippines (Savary et al. 2012). Since the evaluations revealed a reasonably good performance across the region, the parameters of the original EPIRICE were used in our study. The EPIRICE model generates infection risk information on selected rice diseases using weather and cultural data, such as the transplanting date. Because of its broad genericity and simplicity, but sound infection algorithms, EPIRICE has been adopted in many modeling-based studies worldwide (Bunsri 2017; Duku et al. 2016; Kim et al. 2015; Kim and Cho 2016; Sittisak et al. 2017).

To translate the climate information from the tercile SCFs into potential disease risks and because EPIRICE is a daily time-step model, downscaled daily weather realizations were needed. We used the existing downscaling method "the predict-WTD", a parametric method based on a conditional stochastic weather generator that was developed by the International Research Institute for Climate and Society (Ines and Han 2014), to downscale the tercile climate information. The tercile-based SCF was first converted into a theoretical forecast cumulative probability curve. The deviates for each percentile were then converted into rainfall amount, frequency, or intensity to downscale the "full" distribution of the probabilistic SCFs. The seasonal deviates were then disaggregated on a monthly basis and used to constrain the downscaling of the forecast realizations at different percentile values of the theoretical forecast curve (Han and Ines 2017).

In this study, we developed a decision support system where the given SCF for precipitation was first disaggregated into daily weather realizations using the predictWTD method. Temperatures (maximum and minimum) were also generated conditioned on rainfall occurrence in the predictWTD method. EPIRICE was then run with the downscaled weather inputs. Finally, the EPIRICE outputs were visualized as the relative disease risk compared to that of the previous year and the 30-year-climatological average, which generated a seasonal disease prediction. The resulting potential risks for the selected fungal and bacterial diseases of rice can potentially be used to enable decision-makers to develop informed farm management strategies. Such strategic decisions include not only short-term ones (e.g., the selection of resistant cultivars against the high-risk profile diseases, the seasonal scheduling of chemical and cultural control methods, or identifying the ideal time to plant crops to minimize crop losses due to diseases) but also decisions that do not directly pertain to disease management but have numerous crop health consequences (e.g., choices involved in the type of crop establishment, crop rotation, or cropping systems).

12.2.4 Evaluation of the Decision Support System Using an SCF Hindcast

We evaluated the potential applicability of using the APCC PMME SCFs and the decision support system to formulate seasonal disease predictions. Briefly, we downscaled the APCC PMME hindcast grid data over the Pili weather station using the parametric predictWTD method and incorporated these data into the EPIRICE model to produce seasonal disease risk information. This information was then directly compared with the EPIRICE simulation results using the corresponding observed weather data from the Pili weather station. The APCC PMME data used in this study represents 3-monthly, tercile probabilities from a composite of ensemble forecasts from ten independent GCMs. We selected the PMME hindcast for the DJF season as it not only showed the highest hit rate among the seasons (Fig. 12.4) but it also represented the peak DS planting period in the Bicol region. EPIRICE was simulated to produce disease risks for the rice blast, sheath blight, bacterial blight, and brown spot diseases, using both the downscaled PMME hindcast and the observed weather data of the Pili station. The temporal correlation coefficients and root mean square errors were calculated between the two time series of simulated disease severities.

Figure 12.5 shows the graphical comparison between the EPIRICE simulation results for the rice blast and sheath blight diseases using both the downscaled PMME hindcast and the observed Pili weather data for the period of 1983–2005. The time series graphs demonstrated that the PMME hindcast-based disease risk simulations showed good temporal correlations with the observed weather-based data, indicating relatively high predictive skills not only for the climate forecasts but also for their consequences on rice disease infection risks. Statistical analyses resulted in smaller normalized root mean square errors and significant temporal correlation coefficients between the results from the observed weather and the downscaled PMME hindcast data, in comparison to the ones from the observed weather and the climatological data (Table 12.2). In all simulated diseases, reasonably higher temporal correlation coefficients were obtained using the PMME hindcast than the climatological data. This indicated that the interannual variability of the infection risks of the selected diseases depended strongly on precipitation in DJF, and, therefore, the relatively high predictability of the precipitation in the APCC PMME hindcast resulted in the potential predictability of rice disease risks during the DS. The

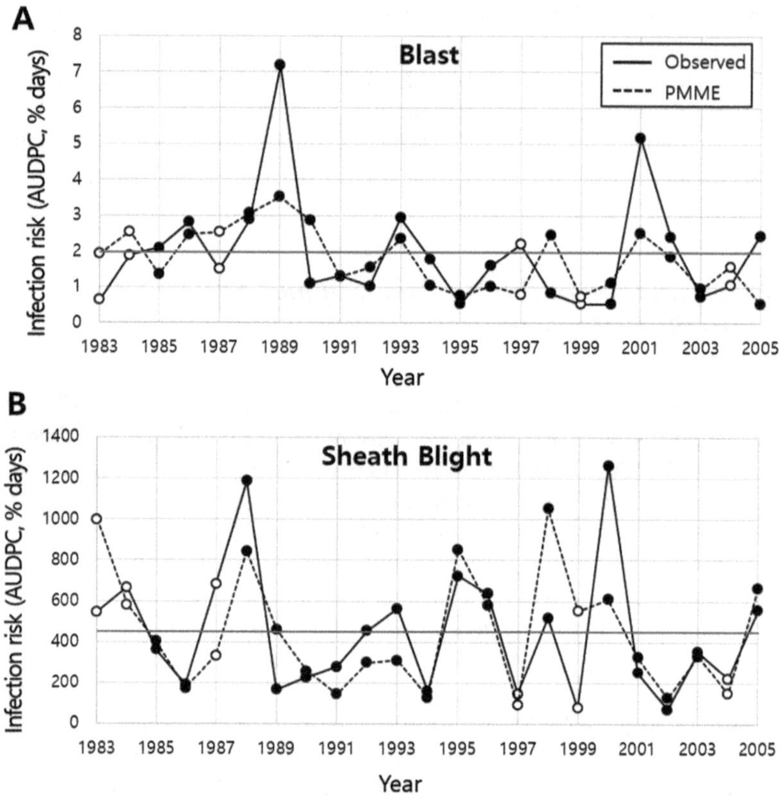

Fig. 12.5 Time series graphs illustrating the EPIRICE simulation results for rice blast (**a**) and sheath blight (**b**) diseases, both using the downscaled APEC Climate Center probabilistic multi-model ensemble (PMME) and observed weather data from the Pili weather station (Observed) for the period of 1983–2005. Black circles indicate the "hit" years, i.e., years when the PMME tercile values matched the observed tercile values. Open circles indicate the "no hit" years. AUDPC: area under the disease progress curve, calculated by accumulating the daily disease severity for the entire growing season (% days)

Table 12.2 Normalized root mean square errors and temporal correlation coefficients between the EPIRICE simulated results for four diseases using observed weather data (obs), downscaled probabilistic multi-model ensemble hindcast (mme), and 30 years of climatological data (clim)

		Blast	Sheath blight	Bacterial blight	Brown spot
Normalized root mean square error	obs & mme	0.66	0.60	0.51	0.42
	obs & clim	0.78	0.70	0.53	0.41
Temporal correlation coefficient	obs & mme	0.5[a]	0.60[a]	0.54[a]	0.40
	obs & clim	−0.04	−0.01	−0.11	0.06

[a]Indicates the correlation coefficients with 99% significance

comparison between the results from the PMME hindcast and those of the climatological data also indicated that the prediction using the PMME hindcast may have some benefits over using the raw climatological data.

12.2.5 Dissemination of the Seasonal Disease Prediction in the Target Area

The Seasonal Climate Forecast and Extension Advisory or CLEA was developed by the Bicol Agri-Water Project (2012–2017) and funded by the USAID for the Bicol region in the Philippines. It is a two-page agro-met bulletin and an easy-to-read reference that was produced to enhance the resiliency of farmers and the ability of local decision-makers to adapt to weather variability and climate changes. Essentially, it serves to help farmers adapt to the variable rainfall patterns over the upcoming season. As a dissemination medium for climate and related agricultural information, the CLEA contains SCFs over the following 6 months, historical rainfall data collected from the rain gauge stations, the ENSO forecast, a tropical cyclone forecast from PAGASA, the results of crop and disease simulations using downscaled SCFs and agricultural models, extensive advisories on ideal times for planting and harvesting, the best crops or varieties to plant, and crop and water management practices recommended by the respective Municipal Agricultural Officer and DA-RFO5, as suitable for the predicted climate and weather conditions. To know and understand the benefits of CLEA, farmers' meetings were organized every month. This motivated farmers to adopt CLEA and benefit from it. It was also in these meetings where the project team got positive feedback that the CLEA has become the basis for their current practices. These were based from farmers' recorded reports containing details about the adopted decision tool and revealing their production performance (Detailed information may refer to the BAWP final report in the USAID website, https://www.usaid.gov).

The relative disease risk levels, compared to the average risk over the past 30 years, for the four major rice diseases were presented as the "Seasonal Disease Outlook" in the CLEA, as shown in Fig. 12.6. The seasonal disease outlook presented the relative disease risk (%) for each disease, as simulated using the given SCFs, which agricultural stakeholders can use to estimate the extent of disease risk abnormality (either lower or higher compared to the risk of the previous year or the average risk from the 30-year climatological data) for the upcoming cropping season. The seasonal disease risk information in the CLEA was provided to farmers before rice planting, i.e., when most important strategic decisions are made, to enable the decision-makers to make informed management decisions. Such decisions include which diseases to prioritize for disease management with limited resources, the volumes of agrochemicals that need to be stocked, which crops or varieties to plant, and which cropping systems to adopt under the anticipated disease pressure. Longer lead times of seasonal time scales are necessary because the

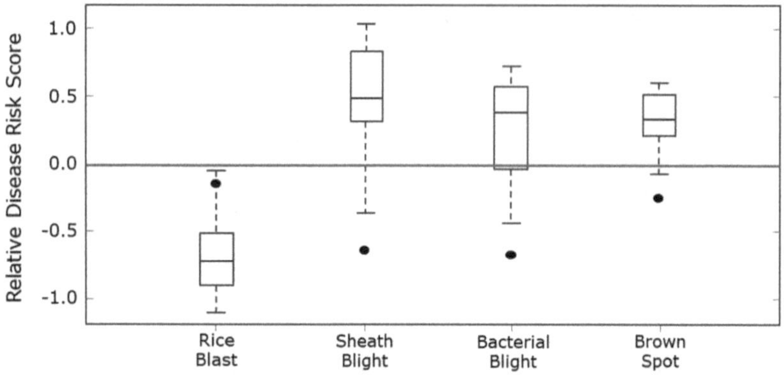

Fig. 12.6 Rice disease risk simulated by EPIRICE for four rice diseases based on the 2005–2006 DJF probabilistic multi-model ensemble seasonal climate forecast hindcast. The horizontal gray line at 0 represents the averaged 30-year climatological risk level for each disease. Black dots indicate the risk score of the previous year, which was also relative to the averaged 30-year climatological risk level. Box boundaries indicate the 25th and 75th percentiles, the line within the box marks the median, and whiskers below and above the box indicate the 10th and 90th percentiles. The relative disease risk scores for each disease simulated with the given SCFs provide a seasonal disease outlook, which agricultural stakeholders can use to estimate the extent of disease risk abnormality (either lower or higher compared to the risk in the previous year or the average risk over the past 30 years) for the upcoming cropping season

majority of efficient counteracting practices for most diseases require an extended preparation time. For example, a lead time of at least 3 months is required to plan and implement proper disease control management actions, such as synchronous planting, replacement with resistant varieties, and changing cropping patterns to including a fallow period.

The information provided in the CLEA can be used to offer recommendations or to assist in the decision-making processes. Since these methods are not designed to identify an "optimal" solution, model-based recommendations to stakeholders should be stated with caution and the role of uncertainty should be emphasized (Han et al. 2017). It is important to use the decision support system and interpret its outputs appropriately. For example, extension workers should interpret the simulated disease risk in relative terms, not by deterministic values. Such interpretations can be achieved by fully understanding the concept of probabilistic SCFs and thus the nature of the resulting disease risks as well. Suggestions and recommendations can then be made to actual decision-makers (e.g., farmers).

12.3 Summary and Conclusions

Herein, we have presented the findings of a case study in Bicol that successfully demonstrated a developmental framework for the application of SCFs to all levels of agricultural decision-making, using a disease epidemiological model to translate

SCFs into agriculturally relevant information. The sequential array of implementation strategies applied in the case study started from a baseline study on disease epidemics in the target area. The selection criteria of the diseases used in the study were based on their economic importance and climatic sensitivity, and rice blast, sheath blight, bacterial leaf blight, and brown spot were identified as the target diseases for use in the seasonal disease predictions. Since these seasonal disease predictions used SCFs, we examined the prediction skill of the APCC PMME SCFs to determine the potential applicability of the SCFs to the target area. RPSS and hit rate analyses of the PMME SCFs against the two different observation datasets revealed that they had a high skill mostly during the DS, in which the DJF season showed the highest hit rate (73.9%). We then developed a decision support system that linked the PMME, particularly for the DS, with a rice disease epidemiological model (EPIRICE) to formulate predictions on the four selected rice diseases. The predictWTD downscaling method was selected to temporally disaggregate the PMME SCFs into daily weather data in the decision support system. The potential applicability of the decision support system was then evaluated using the APCC PMME hindcast from 1983 to 2005. This evaluation was necessary for credibility and to build trust in potential users with regard to the seasonal disease predictions produced by the decision support system. Overall, the relatively high predictability of the precipitation in the APCC PMME hindcast indicated the potential predictability of rice disease infection risks during the DS, using climate information. Lastly, the dissemination of these seasonal disease predictions was demonstrated in the target area through an existing agro-met bulletin called CLEA. The relative disease risk levels for four major rice diseases, compared to those of the previous and the averaged 30-year climatological risks, were presented in CLEA. Using this information, agricultural stakeholders were able to estimate the extent of disease risk abnormality for the upcoming cropping season and thus make informed decisions regarding their disease control efforts.

To broaden the applicability of the findings of the present study to other areas of the Philippines or other developing countries, nevertheless, the following gaps need to be addressed. (1) Not only the disease risk information but information on potential yield losses due to the estimated disease incidences, and their economic values, should be presented. Such data would further "contextualize" the translated information and, therefore, be appropriately based on the needs and capacities of the potential users. To achieve this, extensive data collections would be required in the fields, followed by the development of crop-disease-coupled models for more realistic simulations, based on the given SCFs. (2) The agricultural models used in the decision support system should be subjected to multiple validation processes. In the present case study, the lack of available ground truth data limited the potential for validation. In addition, a number of downscaling approaches should be tested to translate the PMME SCF information into daily scale input data that are accessible to agricultural models. In this study, we only used one downscaling approach which may not have represented the full value of the probabilistic nature of the PMME SCFs. (3) To produce the desired climate information products to be able to support the informed decisions of agricultural stakeholders, it is crucial to guarantee productive coordination between the information providers and users. Productive coordination will increase trust and legitimacy between the two entities and thus facilitate the adoption of the final information generated by the decision support system (Guston 2001; Kirchhoff et al. 2013).

In terms of some broader recommendations to enhance the communication and uptake of the SCF and its translated agricultural information, our interactions with farmers, extension workers, and government officials highlighted that: (1) Forecast information must be translated into a form that is understandable to a range of potential users and thus scientific jargon should be avoided; (2) Forecasts and advisories must be translated into local languages to allow participation from a broad sector of the community; (3) Forecasts and advisories must be localized to ensure their relevance to the cropping calendar for that region; (4) Where possible, forecasts and advisories need to be integrated with agricultural-economic models to provide some economic estimates of the benefits of using such forecasts; and (5) Understanding rural livelihoods and associated agricultural production systems is crucial to translate the SCFs into actionable information.

The value of improved SCFs for the agricultural sector depends on a wide range of complex and interrelated factors. These include forecast accuracy, e.g., accuracy at relevant spatial resolutions and lead times; forecast adoption rates; the attitudes of farmers to risks; and the actual seasonal conditions that are experienced. Therefore, no matter how skilled a climate forecast is, it is not possible for farmers to eliminate all impacts of climate on production, and any actions that they take to mitigate the risks will cost money, as will any decisions they make based on incorrect forecasts. It is unlikely that SCFs will ever achieve complete certainty in forecasting because of the many chaotic and non-deterministic features of climate systems. In fact, due to the anxiety of users regarding the predictability, many systems that have been developed still only use yield estimates from past periods, rather than using actual yield estimates to support informed decision-making. Nevertheless, through the advances in GCM modeling, state-of-the-art downscaling techniques, and the wide application of big-data analyses and artificial intelligence technologies in the climate and agricultural research fields, it is expected that the use of SCFs in agricultural decision-making will increase in the future. To develop comprehensive developmental frameworks for climate services, it is also necessary to consider and simultaneously achieve all three of the key prerequisites for climate services: credibility, salience, and legitimacy (Cash et al. 2003).

Acknowledgment This work was supported by the APEC Climate Center. We are grateful to the partners of the USAID-funded Bicol Agri-Water Project for their sincere support for this collaborative work, especially Dr. Eunjin Han and Dr. Amor Ines for their kind assistance with the predict-WTD downscaling tool within the Climate-Agriculture-Modeling and Decision Tool.

References

Ahn JB, Kim HJ (2013) Improvement of one-month lead predictability of the wintertime AO using a realistically varying solar constant for a CGCM. Meteorol Appl 21:415–418

Boer GJ (2005) An evolving seasonal forecasting system using Bayes' theorem. Atmos Ocean 43:129–143

Bunsri T (2017) Simulation of severity of rice blast disease in Prachin Buri using plant disease epidemiological model: simulation of rice blast disease. In: Proceedings of the 22nd annual meeting in mathematics (AMM 2017)

Canal N, Deudon O, Le Bris X, Gate P, Pigeon G, Regimbeau M, Calvet J-C (2017) Anticipation of the winter wheat growth based on seasonal weather forecasts over France. Meteorol Appl 24:432–443

Cantelaube P, Terres J-M (2005) Seasonal weather forecasts for crop yield modelling in Europe. Tellus A: Dynamic Meteorol Oceanogr 57:476–487

Capa-Morocho M, Ines AV, Baethgen WE, Rodríguez-Fonseca B, Han E, Ruiz-Ramos M (2016) Crop yield outlooks in the Iberian Peninsula: connecting seasonal climate forecasts with crop simulation models. Agric Syst 149:75–87

Cash DW, Clark WC, Alcock F, Dickson NM, Eckley N, Guston DH et al (2003) Knowledge systems for sustainable development. Proc Natl Acad Sci 100(14):8086–8091

Challinor A (2009) Towards the development of adaptation options using climate and crop yield forecasting at seasonal to multi-decadal timescales. Environ Sci Pol 12:453–465

Challinor AJ, Slingo JM, Wheeler TR, Doblas-Reyes FJ (2005) Probabilistic simulations of crop yield over western India using the DEMETER seasonal hindcast ensembles. Tellus A: Dynamic Meteorol Oceanogr 57(3):498–512

CountryStat Philippines (2013) Regional profile: Bicol. http://countrystat.psa.gov.ph/?cont=16&r=5. Accessed 20 Dec 2016)

Duku C, Sparks AH, Zwart SJ (2016) Spatial modelling of rice yield losses in Tanzania due to bacterial leaf blight and leaf blast in a changing climate. Clim Chang 135(3–4):569–583

Epstein ES (1969) A scoring system for probability forecasts of ranked categories. J Appl Meteorol 8:985–987

FAO (2015) The impact of natural hazards and disasters on agriculture and food and nutrition security: a call for action to build resilient livelihoods. (online) Available at: http://www.fao.org/3/a-i4434e.pdf

Goddard L, Mason SJ, Zebiak SE, Ropelewski CF, Basher R, Cane MA (2001) Current approaches to seasonal to interannual climate predictions. Int J Climatol 21(9):1111–1152

Guston DH (2001) Boundary organizations in environmental policy and science: anintroduction. Sci Technol Hum Values 26:399–408

Han E, Ines AV (2017) Downscaling probabilistic seasonal climate forecasts for decision support in agriculture: a comparison of parametric and non-parametric approach. Climate Risk Manage 18:51–65

Han E, Ines AV, Baethgen WE (2017) Climate-Agriculture-Modeling and Decision Tool (CAMDT): A software framework for climate risk management in agriculture. Environ Modell Softw 95:102–114

Han E, Baethgen WE, Ines AV, Mer F, Souza JS, Berterretche M, Atunez G, Barreira C (2018) SIMAGRI: An agro-climate decision support tool. Comput Electron Agric. https://doi.org/10.1016/j.compag.2018.06.034

Hansen JW, Indeje M (2004) Linking dynamic seasonal climate forecasts with crop simulation for maize yield prediction in semi-arid Kenya. Agric For Meteorol 125(1–2):143–157

Hansen JW, Mason SJ, Sun L, Tall A (2011) Review of seasonal climate forecasting for agriculture in sub-Saharan Africa. Exp Agric 47(2):205–240

Hossain M, Gascon F, Revilla I (1995) Constraints to growth in rice in the Philippines. J Agric Econ Dev 33:1–2

Iizumi T, Shin Y, Kim W, Kim M, Choi J (2018) Global crop yield forecasting using seasonal climate information from a multi-model ensemble. Clim Serv 11:13–23

Ines A, Han E (2014) Predict WTD: a temporal downscaling tool for seasonal climate forecast. IRI/Columbia University, New York

IPCC (2014) In: Core Writing Team, Pachauri RK, Meyer LA (eds) Climate change 2014: synthesis report. Contribution of working groups I, II and III to the fifth assessment report of the intergovernmental panel on climate change. IPCC, Geneva

IPCC (Intergovernmental Panel on Climate Change) (2012) In: Field CB, Barros V, Stocker TF, Qin D, Dokken DJ, Ebi KL, Mastrandrea MD, Mach KJ, Plattner G-K, Allen SK, Tignor M, Midgley PM (eds) Managing the risks of extreme events and disasters to advance climate change adaptation. A special report of working groups I and II of the intergovernmental panel on climate change. Cambridge University Press, Cambridge/New York

Janowiak JE, Xie P (1999) CAMS_OPI: a global satellite-raingauge merged product for real-time precipitation monitoring applications. J Clim 12:3335–3342

Kanamitsu M, Kumar A, Juang HMH, Schemm JK, Wang W, Yang F et al (2002) NCEP dynamical seasonal forecast system 2000. Bull Am Meteorol Soc 83(7):1019–1038

Kang HS, Boo KO, Cho CH (2011) Introduction to KMA-Met office joint seasonal forecasting system and evaluation of its hindcast ensemble simulations. NOAA/NWS Science and Technology Infusion Climate Bulletin

Kharin VV, Zwiers FW (2003) Improved seasonal probability forecast. J Clim 16:1684–1701

Kim KH, Cho J (2016) Predicting potential epidemics of rice diseases in Korea using multi-model ensembles for assessment of climate change impacts with uncertainty information. Clim Chang 134(1–2):327–339

Kim KH, Cho J, Lee YH, Lee WS (2015) Predicting potential epidemics of rice leaf blast and sheath blight in South Korea under the RCP 4.5 and RCP 8.5 climate change scenarios using a rice disease epidemiology model, EPIRICE. Agric For Meteorol 203:191–207

Kirchhoff CJ, Lemos MC, Dessai S (2013) Actionable knowledge for environmental decision making: broadening the usability of climate science. Annu Rev Environ Resour 38:393–414

Koide N, Robertson AW, Ines AV, Qian JH, DeWitt DG, Lucero A (2013) Prediction of rice production in the Philippines using seasonal climate forecasts. J Appl Meteorol Climatol 52(3):552–569

Lim EP, Hendon HH, Langford S, Alves O (2012) Improvements in POAMA2 for the prediction of major climate drivers and south eastern Australian rainfall. CAWCR technical report no. 051

Liou CS, Chen JH, Terng CT, Wang FJ, Fong CT, Rosmond TE, Kuo HC, Shiao CH, Cheng MD (1997) The second generation global forecast system at the central weather bureau in Taiwan. Weather Forecast 12:653–663

Lyon B, Cristi H, Verceles ER, Hilario FD, Abastillas R (2006) Seasonal reversal of the ENSO rainfall signal in the Philippines. Geophysical Res Letters 33(24):L24710

Martins MA, Tomasella J, Rodriguez DA, Alvalá RC, Giarolla A, Garofolo LL et al (2018) Improving drought management in the Brazilian semiarid through crop forecasting. Agric Syst 160:21–30

Mason SJ, Goddard L (2001) Probabilistic precipitation anomalies associated with EN SO. Bull Am Meteorol Soc 82(4):619–638

Meinke H, Stone RC (2005) Seasonal and inter-annual climate forecasting: the new tool for increasing preparedness to climate variability and change in agricultural planning and operation. Climate Change 70:221–253

Merryfield WJ et al (2013) The Canadian seasonal to interannual prediction system. Part I: models and initialization. Mon Weather Rev 141:2910–2945

Molod A et al (2012) The GEOS-5 atmospheric general circulation model: mean climate and development from MERRA to Fortuna. Technical report series on global modeling and data assimilation, p 28

Murphy AH (1969) On the ranked probability skill score. J Appl Meteorol 8:988–989

Murphy AH (1971) A note on the ranked probability skill score. J Appl Meteorol 10:155–156

Philippine Department of Agriculture, Regional Office No. 5 (2014) Bicol annual report 2014. http://bicol.da.gov.ph/index.php/reports-documentation/1207-da-bicol-annual-report-2014

Power SB, Plummer N, Alford P (2007) Making climate models more useful. Aust J Agric Res 58:945–951

Rosenzweig C, Jones JW, Hatfield JL, Ruane AC, Boote KJ, Thorburn P et al (2013) The agricultural model intercomparison and improvement project (AgMIP): protocols and pilot studies. Agric For Meteorol 170:166–182

Savary S, Nelson A, Willocquet L, Pangga I, Aunario J (2012) Modeling and mapping potential epidemics of rice diseases globally. Crop Prot 34:6–17

Scinocca JF, Mcfarlane NA, Lazare M, Li J, Plummer D (2008) The CCCma third generation AGCM and its extension into the middle atmosphere. Atmos Chem Phys 8:7055–7074

Sittisak I, Saruda H, Angkool W, Thidarat B (2017) Numerical solution of the differential equation for simulation of the rice blast disease. J Appl Sci Environ Manag 21(7):1272–1275

Syktus J, McKeon G, Flood N, Smith I, Goddard L (2003) Evaluation of a dynamical seasonal climate forecast system for Queensland. In: Science for drought, Proceedings of the National Drought Forum, Brisbane, April 2003. Queensland Department of Primary Industries. Ed. R Stone and I Partridge, pp 160–173

Takaya Y, Yasuda T, Ose T, Nakaegawa T (2010) Predictability of the mean location of typhoon formation in a seasonal prediction experiment with a coupled general circulation model. J Meteorol Soc Japan 88:799–812

Trosnikov IV, Kaznacheeva VD, Kiktev DB, Tolstikh MA (2005) Assessment of potential predictability of meteorological variables in dynamical seasonal modeling of atmospheric circulation on the basis of semi-Lagrangian model SL-AV. Russian Meteorol Hydrol 12:5–17

Chapter 13
Development, Validation, and Dissemination of a Decision Support System for Rainfed Rice Farming in Southeast Asia: A Case Study in Indonesia

Keiichi Hayashi, Lizzida Llorca, and Iris Bugayong

Abstract Rice farmers in rainfed areas face various constraints due to unfavorable environment caused by weather-related uncertainties. The weather-rice-nutrient integrated decision support system (WeRise), a seasonal climate prediction-based online application tool, was developed through the IRRI-Japan Collaborative Research Project (IJCRP) to improve the current situation of rainfed farmers. WeRise provides critical crop management advisory such as optimum sowing timing. The predictability of WeRise was evaluated through hind cast analysis and on-farm field validation in Indonesia. Results from hind cast analysis showed a high predictability of WeRise as demonstrated through a good model fit measured by normalized root mean square errors (RMSEn). Two groups of farmers using Ciherang variety in their cropping obtained different grain yields. The group who used WeRise advisories obtained significantly better grain yield than the group who did not. This also demonstrates another measure of the predictability of WeRise. In addition to technology development and validation, a technology transfer pathway was developed by IJCRP to identify stakeholders and their roles in the uptake and dissemination of WeRise. It involves stakeholders from the national and local governments. Capacity building of national government agencies who could be in-charge of localization/further development, operation, and maintenance of WeRise is among the components of the pathway. The integration of WeRise into the policy, research and development programs was also found to be crucial to facilitate technology transfer. Moreover, policy advocacy should be also conducted by IJCRP within the project implementation period for solving inherent problems including

K. Hayashi (✉)
Japan International Research Center for Agricultural Sciences, Tsukuba, Ibaraki, Japan
e-mail: khayash@affrc.go.jp

L. Llorca · I. Bugayong
International Rice Research Institute, University of the Philippines Los Baños Compound, Laguna, Philippines

T. Iizumi et al. (eds.), *Adaptation to Climate Change in Agriculture*,
https://doi.org/10.1007/978-981-13-9235-1_13

data sharing and coordination among the national and local agencies. Monitoring and evaluation (M&E) and post-technology transfer support are also key activities of IJCRP. While Southeast Asian countries may have different government structures, system, and information flow, the WeRise technology transfer pathway developed for Indonesia could serve as a guide for other countries and regions.

13.1 Introduction

Rice is one of the staple crops that feeds more than 3 billion people of the world. Its production needs to be enhanced due to the population increase in the future. On the other hand, rice production is one of the contributors of greenhouse gas emissions such as methane. Environmental pollution is also one of the areas of concern because rice production causes direct and indirect degradation of the ecosystem. Furthermore, rice production requires large amounts of water and land. Competition of different sectors over finite resources is becoming an issue due to climate change. These problems can be solved through a holistic approach. Hence, it is imperative to use relevant technologies to deal with existing problems and concerns, which enable farmers to produce more rice to feed the increasing population in a sustainable manner.

Irrigated rice accounts for 58% of the total global rice production area and 75% of the global rice production which can be largely attributed to a favorable water supply through irrigation system. To meet future demand, an average increase in grain yield of 0.6 ton per hectare per year is needed over the next decade (GRiSP 2013). However, this may be difficult to achieve if rice supply depends only on irrigated rice production due to existing and potential constraints in natural resources like water and land (Wassmann et al. 2009). Closing the gap between the attainable and potential grain yield of irrigated rice varieties may not be easy as the growth rate of rice grain yield has not been increasing since the 1960s (FAO 2018). Thus, enhancing rice production in other ecosystems like rainfed areas is one way to supplement irrigated rice production.

Rainfed rice areas account for only less than 20% of the global rice production because its average yield is only around 2 t ha^{-1} compared to 5 t ha^{-1} average yield of irrigated rice (GRiSP 2013). However, its production area is around 40% of the total global production area indicating its potential contribution to future rice supply (Haefele et al. 2014). A study which looked into the spatial and temporal variations in productivity of rainfed and irrigated rice areas in the Philippines indicated that both ecosystems were achieving similar productivity levels, and the yield gap between these ecosystems can be closed through facilitating the adoption and diffusion of relevant technologies (Mariano et al. 2010). Seasonal climate predictions (SCP) are useful for farmers' decision-making. There are several studies on the application of SCP on uplands crops (e.g., Hansen et al. 2009) that are related to this. The application of SCP in rice production seems useful, but available information is limited in Asia (Hayashi et al. 2018).

The weather-rice-nutrient integrated decision support system (WeRise) was developed through the IRRI-Japan Collaborative Research Project funded by the Ministry of Agriculture, Forestry and Fisheries of Japan. WeRise is a seasonal climate prediction-based decision support system for rainfed rice areas. It is designed as a web-based tool (Hayashi and Llorca 2016) which guides rainfed rice farmers to use an appropriate variety with relevant information for cropping. The purpose of this paper is to evaluate the applicability of WeRise as a decision support system for rainfed rice farming and to propose a technology transfer pathway in the context of Indonesia.

13.2 Constraints in Rainfed Rice Production

Rice production needs larger amounts of water than upland crops. To produce 1 kg of rice grain, 2.5 m^3 of water is required (Bouman et al. 2007). This is equivalent to 12,500 m^3/ha of water to harvest 5 t ha^{-1} of rice grain. Hence, water management is crucial to obtain high grain yield at the end of every season. However, water supply in rainfed rice areas is uncertain because it comes from rainfall which has large annual fluctuation at the start and end of the season, temporal distribution, as well as total amount. Due to this, rainfed rice farming is often exposed to high risks, such as droughts and floods during the cropping season, which directly cause substantial yield reduction. Another constraint in rainfed rice areas is poor nutrient supply for rice growth due to the type of soil distributed in the areas. Poor and very poor soils account for majority of rainfed rice areas in Asia (54% of the total rice cultivated area) and Africa (80%) (Haefele et al. 2014). Soils in tropical and sub-tropical areas contain dominantly highly weathered clay minerals such as kaolinite that has low water and nutrient holding capacity. This type of soil coincides with rainfed rice areas. Because soil fertility is one of the crucial factors for better grain yield, fertilizer application is essential, especially in rainfed rice areas where the soil has low nutrient supplying capacity. Unfortunately, fertilizer application cannot be properly practiced by rice farmers because it is difficult to apply fertilizer without a reliable water supply. Farmers often experience that there is no enough water in the paddy field when fertilizer application is scheduled. Sometimes a downpour may occur right after fertilizer application that reduces nutrient absorption by the plant due to flushing of the fertilizer. Fertilizer is an expensive investment for small-scale farmers. With their limited economic capacity for the investment, they are reluctant to apply fertilizer due to high risk and low return.

Rainfed rice production relies on rainfall for its water supply; hence, local farmers start cropping as soon as the rainy season arrives. However, there is a lag between the onset of rainy season and sowing. This is farmers' practice as a form of local risk management in Southeast Asia. Surveys were conducted in Indonesia and Lao PDR to understand their local practices in determining the sowing criteria at the beginning of a cropping season. According to the results, the average cumulative rainfall and number of days from the first rain event of the season were 139 mm and 100 days

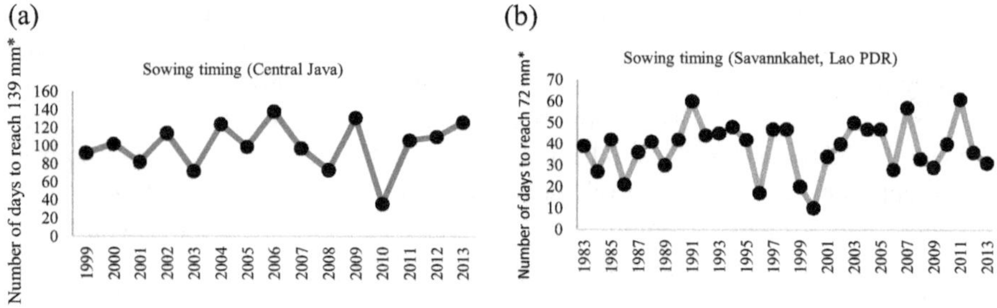

Fig. 13.1 Fluctuation of sowing timing in (**a**) Central Java, Indonesia and (**b**) Savannakhet, Lao PDR. Fluctuation of sowing timing is represented by the number of days required to achieve a cumulative rainfall of (**a**) 139 mm and (**b**) 72 mm from August 1st. (The data are based on Hayashi et al. 2018)

for Central Java, Indonesia, and 72 mm and 38 days for Savannakhet, Lao PDR, respectively. This is a local risk avoidance to enable young seedlings to escape water shortage at the early stage of the season. In this practice, farmers observe the field condition prior to deciding sowing timing. Hence, they need to wait for a period of time depending on the rainfall pattern and amount every year. Accordingly, sowing timing fluctuates as shown in Fig. 13.1.

As Fig. 13.2 shows, farmers in Central Java waited an average of 100 days with a maximum of 132 days and minimum of 36 days prior to starting sowing according to the local criteria of cumulative rainfall amount. In Savannakhet, farmers waited for 38 days on the average with a maximum of 61 days and minimum of 10 days. Usually, cropping season lasts from October to February in Central Java, which is located in the Southern Hemisphere, and June to October in Savannakhet, which is located in the Northern Hemisphere. However, these cropping seasons fluctuate according to the onset of the rainy season as the arrival of Asian monsoon changes every year. Local farmers try to pick a sowing date with the hope of a better probability of success in production despite uncertainties. Figure 13.3 shows the actual grain yield as a function of different sowing timings in the case of Central Java which shows as a wide sowing window spanning almost 2.5 months from October 5th to December 24th. During this window, grain yield fluctuates according to the sowing date which seems to happen by chance. Farmers would never know the result of their harvest until the end because they cannot anticipate weather events during the cropping season.

Drought is also one of the causes of substantial grain yield loss and is considered the most severe abiotic stress (Schiller et al. 2006). When the rainy season arrives as expected and sowing has taken place at the appropriate period of the cropping season, sufficient length of crop growth period could be expected due to the minimum risk of terminal drought. On the other hand, when sowing has taken place late due to the late arrival of the rainy season or erratic rainfall pattern at the early stage of the cropping season, the probability of terminal drought seems to be high due to a short

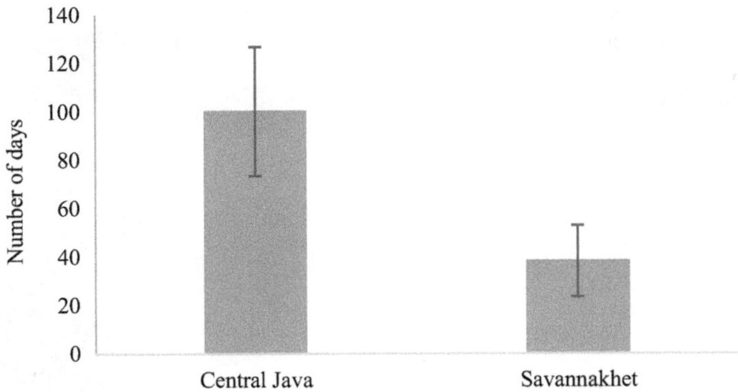

Fig. 13.2 Sowing criteria for local farmers' sowing timing in Southeast Asia

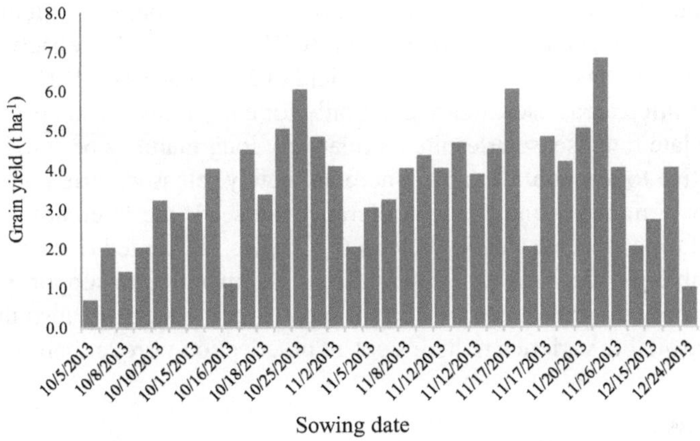

Fig. 13.3 Measured grain yields as function of sowing dates for early rainy season in 2013, Central Java

crop growth period. Furthermore, it is difficult to avoid thoroughly the risk of drought during the cropping season no matter how timely farmers start sowing. Farmers need to sow twice or thrice if drought hits young seedlings at the early stage of the growing season even if sowing has taken place according to the local criteria. In case an early maturing variety is used, the peak of the rainy season could coincide during flowering which may also cause a significant yield loss (Tuong et al. 2000). It is impossible for farmers to anticipate precisely the risks during the cropping season and make an appropriate decision to avoid risks and make outputs better.

One way to overcome the existing constraint in rainfed areas is through appropriate varietal selection which gives rainfed rice farmers some options for risk management. However, the result of a baseline survey in Indonesia showed that the varieties

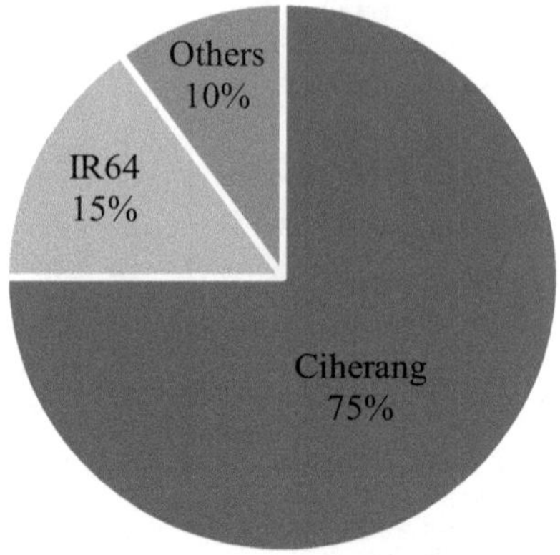

Fig. 13.4 Varieties used by rainfed rice farmers in Central Java. (Data are collected by farmers' interview)

used by rainfed rice farmers were limited and one popular variety was dominantly used for production (Fig. 13.4). Ciherang is one of the improved varieties released by the Indonesian Center for Rice Research (ICRR) in 2000. This variety matures in 116–125 days and has an attainable grain yield of 5–7 t ha^{-1} (BP PADI 2015). Most farmers in rainfed areas use Ciherang not only for early rainy season production but also for the late rainy season despite its relatively long maturity period. Eventually, it is susceptive to terminal drought. There are newly released varieties which have shorter days of maturity and drought tolerance but seeds are often not available for the rainfed farmers because these "tolerant" varieties for rainfed rice production are not marketable for seed growers. Therefore, it is imperative to support rainfed rice farmers not only by improving the current seed supply system but also through utilization of available varieties in their current practice to improve their production.

13.3 Weather-Rice-Nutrient Integrated Decision Support System (WeRise)

Weather forecast is a useful information to guide rainfed rice farmers in their sowing and fertilizer application. However, a successful production in terms of grain yield is not only due to water availability at the time of sowing/fertilizer application but also due to the distribution of water supply throughout the cropping season, especially from panicle initiation to heading during the reproductive stage. Furthermore, weather forecast is not enough to guide the farmers to use locally available varieties like Ciherang properly. Hence, season-long weather information is needed to guide the farmers on optimum sowing timing and appropriate varietal selection.

WeRise is a seasonal climate prediction-based decision support system for rainfed rice production. Developed through statistical downscaling of SINTEX-F ocean-atmosphere coupled general circulation model (GCM developed by Japan Agency for Marine-Earth Science and Technology, JAMSTEC) daily outputs to be an input to ORYZA, a crop growth model, it provides grain yield predictions as a function of sowing dates. The grain yield predictability of WeRise was validated in Central Java, Indonesia (Hayashi et al. 2018).

13.3.1 Predictability of WeRise and Its Limitation

A cumulative distribution function-based downscaling method (CDFDM) is a statistical downscaling method to correct biases in GCM daily output (Iizumi et al. 2010). Figure 13.5 shows the skill of CDFDM for various weather parameters from the outputs of SINTEX-F in the case of Indonesia. The discrepancy between locally observed weather data and outputs of SINTEX-F was minimized in most of the parameters after CDFDM application.

Figure 13.6 shows the result of hindcast analysis to evaluate the predictability of WeRise in Indonesia for different locations and varieties. Observed weather data was used in ORYZA to obtain grain yield retroactively, and the grain yield was compared with the outputs from WeRise. A normalized root mean square error (RMSEn) was employed for this evaluation on grain yields between ORYZA and WeRise. Result for Inpari 39, Inpari 41, and Ciherang shows an RMSEn of 29.9%, 25.4%, and 17.3%, respectively, for North Sumatra and 8.35%, 9.78%, and 28.31%, respectively, for Central Java, which demonstrates a good model fit according to Jamieson et al. (1991). WeRise predictability was also tested through on-farm field trial in early rainy seasons in Indonesia using Ciherang, one of the most popular varieties across the country, and results are shown in Fig. 13.7. Grain yields between two groups of farmers, one group with WeRise advisories and another without advisories, were significantly different. WeRise advisories were given based on the national fertilizer recommendation and showed that grain yield increased after the end of October (Hayashi et al. 2018). Some farmers started before this timeline which resulted in lower yields implying the importance of an optimum sowing

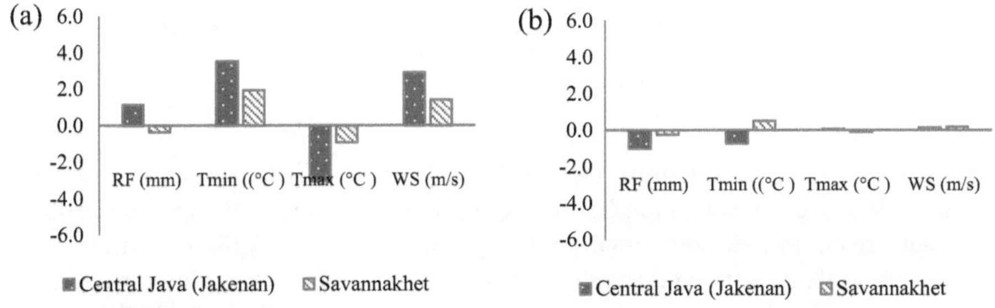

Fig. 13.5 Mean errors for rainfall (RF), minimum air temperature (Tmin), maximum air temperature (Tmax), and wind speed (WS) between (**a**) observed weather data and SINTEX-F outputs without bias correction and (**b**) observed weather data and SINTEX-F outputs with bias correction based on CDFDM

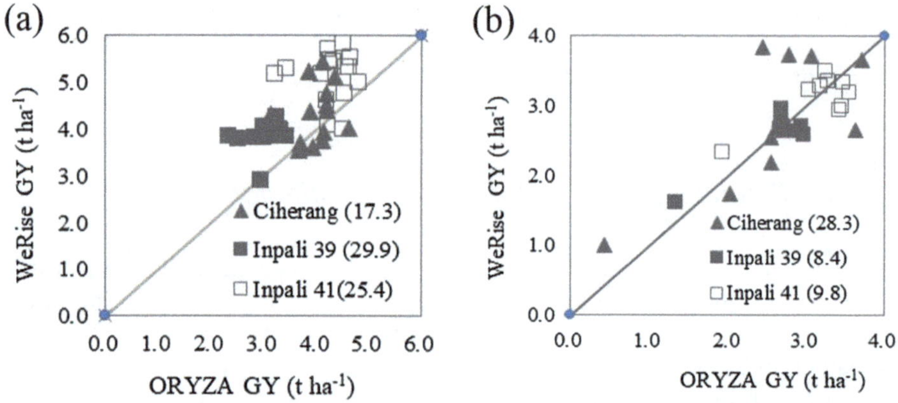

Fig. 13.6 Grain yield comparison between ORYZA and WeRise for (**a**) North Sumatra and (**b**) Central Java. Numbers in parenthesis indicate the value of normalized root mean square errors (RMSEn)

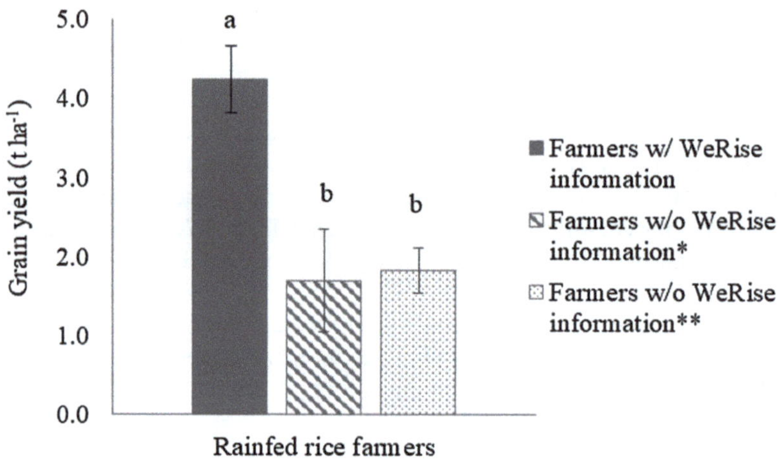

Fig. 13.7 Results of on-farm validation on WeRise predictability. * indicates that farmers have sown earlier than farmers with WeRise advisories; and ** indicates that farmers have sown almost during the same period of WeRise advisories

timing. Meantime, the low grain yield of farmers whose sowing coincided with the period of WeRise advisories implies the importance of proper fertilizer management to benefit from an optimum sowing timing. The strength of WeRise in improving grain yield as demonstrated through these results is promising for rainfed rice farmers. However, CDFDM bias correction is not effective to apply in SINTEX-F outputs when rainfall is above normal (Hayashi et al. 2018). Hence, WeRise can be only applicable for drier years compared to normal year. Furthermore, the limitation of the mathematical model should also be taken into consideration to understand the discrepancy between model and actual grain yields.

13.3.2 Output Information of WeRise

To obtain predicted grain yield, WeRise provides information through two advisories, weather and crop advisories (Fig. 13.8). The weather advisories which help farmers to understand the characteristics of the upcoming weather of the next cropping season include information such as the onset of the rainy season, distribution

Fig. 13.8 Front page of WeRise online application accessible through the IRRI website (http://beta.werise.irri.org/index.php)

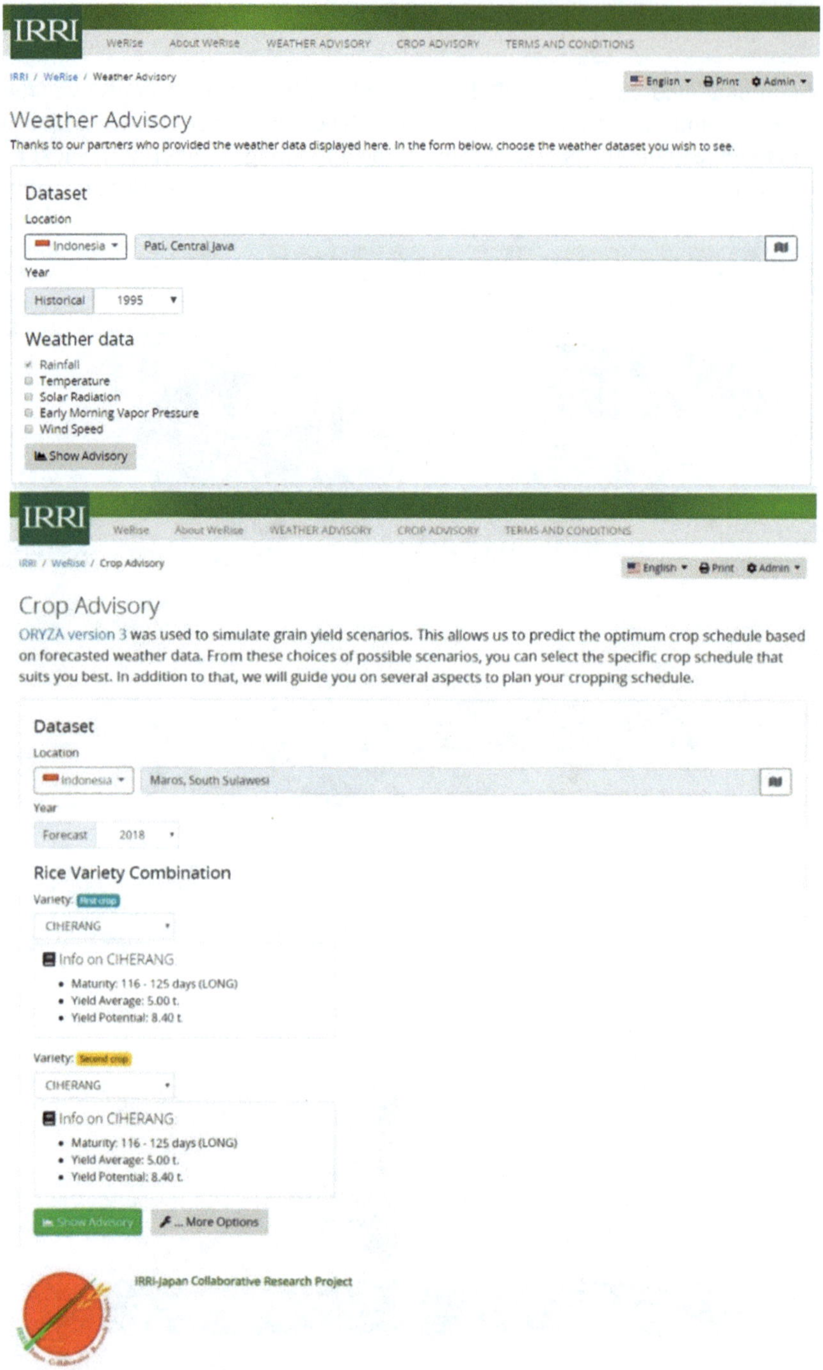

Fig. 13.9 Sample inputs for WeRise weather and crop advisories (http://beta.werise.irri.org/index.php)

of rainfall, period of droughts, and flood risks. In addition to rainfall, the weather advisories include other weather parameters such as maximum and minimum air temperature, solar radiation, early morning vapor pressure, and wind speed. These are required parameters for ORYZA to produce grain yield predictions (Fig. 13.9).

The crop advisories could help farmers in their decision-making through choosing appropriate varieties according to the weather predictions (Fig. 13.9). WeRise enables varietal selection for two rice croppings to demonstrate grain yield as a function of sowing dates. It allows farmers to choose a suitable variety and optimum sowing timing three months before the next cropping season. This allows farmers enough time to prepare seeds and fertilizers, arrange for/plan labor requirements, and make reservations for machinery for land preparation. In general, rainfed rice farmers are reluctant to apply chemical fertilizer because of uncertainties in water availability in the field. An adequate level of surface water is required for chemical fertilizer application to be effectively absorbed by rice. Split fertilizer application through not only basal but also topdressing application is also highly recommended for better grain yield (Peng et al. 1996). In rainfed condition, however, fertilizer application is a high-risk investment for rainfed rice farmers due to uncertainties in water supply; hence, the farmers do not follow the recommendation. WeRise determines the suitable fertilizer application timings according to the predicted weather during critical crop growth stages such as maximum tillering and panicle initiation. This could serve as a guide for rainfed farmers to utilize their finite resource efficiently and improve productivity.

13.4 Dissemination of WeRise and Its Adoption Pathway to Rainfed Rice Farmers

WeRise was developed as a web-based tool to allow many users to access from different locations and obtain critical information for better decision-making. Figure 13.10 shows a conceptual diagram of the background operational system of WeRise. It operates through an integration of databases from ORYZA and CDFDM. Since WeRise is a location-specific decision support system, database development should be done per location. Outputs from SINTEX-F should be corrected for biases with local historical weather data through CDFDM. Predictions in required weather parameters are stored in a database to be used by ORYZA for grain yield predictions as function of sowing dates.

The weather database needs to be updated every year, three months before the cropping season to give rainfed rice farmers enough time to prepare for the upcoming cropping season. Crop database needs to be updated according to the availability of new varieties and farmers' preference. ORYZA calibration should be done to update the information on crop growth including phenology in the database. Crop and weather databases are stored in a webserver and used to generate the WeRise predictions (Hayashi and Llorca 2016).

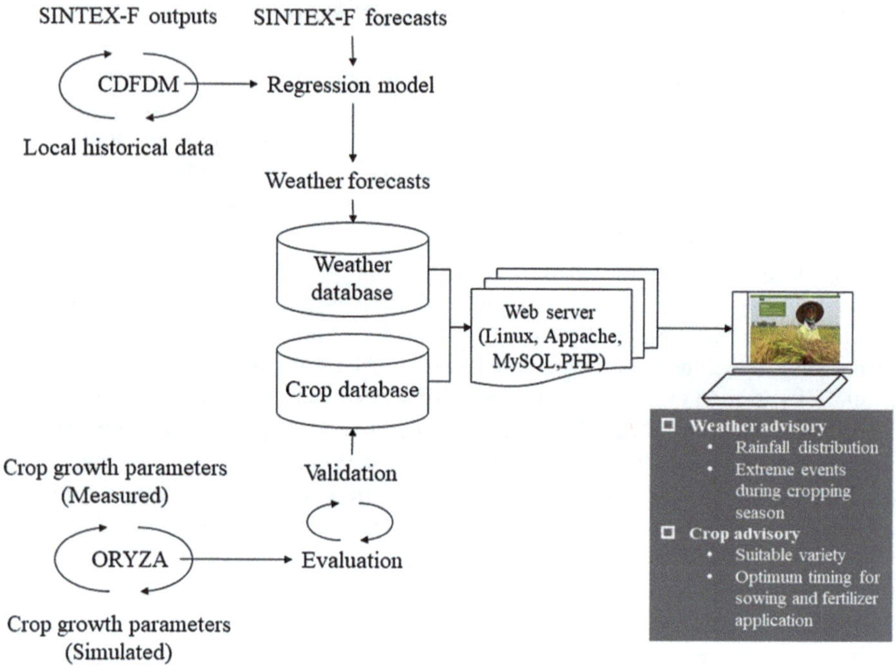

Fig. 13.10 Conceptual diagram of the background operational system and outputs of WeRise

A technology transfer pathway for WeRise in Indonesia was developed to facilitate a smooth transition and sustain project outputs after IJCRP's implementation period (Fig. 13.11). The pathway shows the stakeholders and their potential roles in the uptake and dissemination of WeRise.

In addition to technology development and validation, IJCRP should conduct capacity building of key technology transfer recipients within the project implementation period. These recipients are national government agencies under the Indonesian Agency for Agricultural Research and Development (IAARD) of Indonesia's Ministry of Agriculture (MoA). For localization/further development, operation, and maintenance, these are the Indonesian Agroclimate and Hydrology Research Institute (IAHRI) and the Indonesian Center for Rice Research (ICRR). The Assessment Institute for Agricultural Technology (AIAT) in the rainfed rice producing provinces, in coordination with extension at the local government level, could be in-charge of assessment and information dissemination. Local government extension workers could serve as information intermediaries by disseminating WeRise advisories, collecting feedback from farmers, and communicating feedback to AIAT.

IAHRI, ICRR, and AIAT are being coordinated by the Indonesian Center for Agricultural Land Resources Research and Development (ICALRRD), Indonesian Center for Food Crops Research and Development (ICFORD), and Indonesian Center for Agricultural Technology Assessment and Development (ICATAD), respectively. These coordinating institutions could serve as links to the IAARD director.

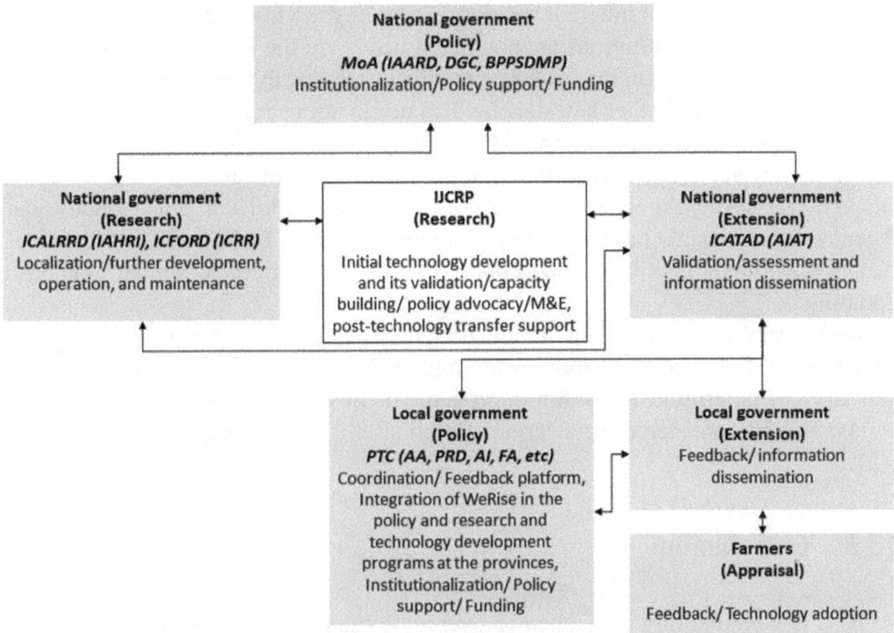

Fig. 13.11 WeRise technology transfer pathway in Indonesia

IAARD and other national government agencies under Indonesia's MoA including the Directorate General of Crops (DGC) and Agency for Agricultural Human Resources Development (BPPSDMP) and local government at the district and provincial levels are key stakeholders for institutionalization, policy support, and funding to sustain WeRise.

Most of the stakeholders in the proposed technology pathway are already performing the proposed roles above as part of the National and Provincial Taskforces for Indonesia's Cropping Calendar Information System (CCIS) Technology locally known as Kalender Tanam or Katam. WeRise could be a special module of Katam specifically targeting rainfed rice areas. IJCRP has been working with most of them for the development and validation of WeRise.

The local government could also help with the integration of WeRise in the policy and research and technology development programs of the provinces through the Provincial Technology Commission (PTC). The PTC is a coordinating platform for AIAT and other stakeholders including agricultural agencies (AA), Provincial Research and Development Agency (PRDA), province-based academic institutions (AI) such as local universities, farmer associations (FA), legislative members, and chambers of trade and commerce. PRDA coordinates and implements research and development activities by itself or with implementing organizations such as universities, at the provincial level for planning purposes. It could thus also serve as a coordinating/feedback platform for WeRise. This would forego the need to create new committees or working groups solely for WeRise which might not be sustainable.

However, PTCs have inherent problems including the frequent change in leadership of its member institutions as incumbent heads are the PTC representatives by default, low participation rate as some members are inactive and do not attend the regular meetings, and other key stakeholders might not be represented (e.g., district level agencies) (Saediman 2015). IJCRP may conduct policy advocacy in this regard. Policy advocacy may also help facilitate data sharing among IAHRI, ICRR, and the Indonesian Agency for Meteorological, Climatological, and Geophysical Agency – Indonesia's national climate center (BMKG). BMKG is also a part of Katam's National Taskforce and supports IAHRI in data collection and database updating.

Within the project implementation period which is until September 30, 2020, IJCRP should also conduct monitoring and evaluation of the technology transfer activities to determine areas for improvement. Post-technology transfer support should also be considered for a certain period after the end of the project.

13.5 Conclusion

WeRise is a technology developed to overcome the existing constraints in rainfed rice areas in Southeast Asia. It is a web-based tool for strategic crop production planning accessible to many users. It enables data-driven decision support for rainfed rice system through science-based weather and crop advisories. However, the challenge is not only to provide technology for the existing problems but also to facilitate adoption by the end-users. A technology transfer pathway identifies the key stakeholders and their roles in the uptake and dissemination WeRise. It is imperative to involve the key stakeholders at the early stage to share the project concept, create a shared sense of ownership on WeRise, and gain their commitment to sustain project outputs and activities after the project implementation period. These activities include localization, operation and maintenance, further development, validation, dissemination, and institutionalization and policy support at the national and local levels. While Southeast Asian countries may have different government structures, system, and information flow, the WeRise technology transfer pathway developed for Indonesia can be a guide for other countries and regions.

Acknowledgment WeRise was developed through the IRRI-Japan Collaborative Research Project (JCRP) funded by the Ministry of Agriculture, Forestry and Fisheries of Japan.

References

BB PADI (2015) Deskripsi Varietas Unggul Baru Padi pp 77

Bouman BAM, Lampayan RM, Tuong TP (2007) Water management in irrigated rice: coping with water scarcity. International Rice Research Institute. p 54

Food and Agriculture Organization of the United Nations (FAO) (2018) Agricultural population. In: FAOSTAT. http://faostat.fao.org/site/291/default.aspx. Browsed on July 11, 2018

GRiSP (Global Rice Science Partnership) (2013) Rice Almanac, 4th edn. International Rice Research Institute, Los Banos

Haefele SM, Nelson A, Hijmans RJ (2014) Soil quality and constraints in global rice production. Geoderma 235–236:250–259

Hansen JW, Mishra A, Rao KPC, Indeje M, Ngugi RK (2009) Potential value of GCM-based seasonal rainfall forecasts for maize management in semi-arid Kenya. Agric Syst 101:80–90

Hayashi K, Llorca L (2016) The weather-rice-nutrient integrated decision support system (WeRise), JIRCAS working report 83

Hayashi K, Llorca L, Rustini S, Setyanto P, Zaini Z (2018) Reducing vulnerability of rainfed agriculture through seasonal climate predictions: a case study on the rainfed rice production in Southeast Asia. Agric Syst 162:66–76

Iizumi T, Nishimori M, Ishigooka Y, Yokozawa M (2010) Introduction to climate change scenario derived by statistical downscaling (in Japanese). J Agric Meteorol 66:131–143. https://doi.org/10.2480/agrmet.66.2.5

Jamieson PD, Porter JR, Wilson DR (1991) A test of the computer simulation model ARCWHEAT 1 on wheat crops grown in New Zealand. Field Crop Res 27:337–350

Mariano MJ, Villano R, Fleming E (2010) Are irrigated farming ecosystems more productive than rainfed farming systems in rice production in the Philippines? Agric Ecosys Envi 139(4):603–610

Peng S, Garcia FV, Laza RC, Sanico AL, Visperas RM, Cassman KG (1996) Increase N-use efficiency using a chlorophyll meter on high-yielding irrigated rice. Field Crops Res 47:243–252

Saediman H (2015) Improving agricultural research coordination at subnational level in Indonesia: an assessment of opportunities for strengthening Provincial Technology Commission. Int J Agric Ext 3(2):123–136

Schiller JM, Chanphengxay MB, Linquist B, Appa Rao S (eds) (2006) Rice in Laos. International Rice Research Institute, Los Baños

Tuong TP, Singh AK, Siopongco JDLC, Wade LJ (2000) Constraints to high yield of dry-seeded rice in the rainy season of a humid tropic environment. Plant Prod Sci 3(2):164–172

Wassmann R, Jagadish SVK, Heuer S, Ismail A, Redona E, Serraj R, Singh RK, Howell G, Pathak H, Sumfleth K (2009) Climate change affecting rice production: the physiological and agronomic basis for possible adaptation strategies. In: Donald L, Sparks DL (eds) Advances in agronomy, vol 101. Academic, Burlington, pp 59–122

Part IV
Potential Ways Moving Forward

Chapter 14
The Roles of Farmers, Scientists, and Extension Staff in Technology Development for Soil Frost Control as an Adaptation to Climate Change in Tokachi, Hokkaido, Japan

Tomoyoshi Hirota and Kazuhiko Kobayashi

Abstract For any measures of agricultural adaptation to climatic change, it is critical that they be adopted by most farmers in the target region. We investigated farmers' wide adoption of soil frost control (SFC) measures against the proliferation of volunteer potatoes (VP) due to climate change in the Tokachi region of Hokkaido, Japan, making use of concepts in the *diffusion of innovations* paradigm. In the initial stage, snowplowing spread as a farmer-invented innovation to suppress VP, but the adopters sometimes encountered undesired outcomes due to insufficient or excessive soil frost. An extension staff brought this problem to the attention of a scientist, who had developed a soil temperature model. With the model, the farmer practice evolved into the SFC technology to attain the optimum soil frost. The technology was implemented in a web-based decision-support system for use by the farmers, which facilitated its wide adoption by farmers across the region. The two critical steps to SFC: the farmer's invention of snowplowing and the scientist's development of soil temperature model, were made with no intention for the use to climate change adaptation, instead their use was found on the farmers' network initially and the network of scientists and extension staffs later.

T. Hirota (✉)
NARO Hokkaido Agricultural Research Center, Sapporo, Japan
e-mail: hirota@affrc.go.jp

K. Kobayashi
The University of Tokyo, Tokyo, Japan

Ibaraki University, Ami, Japan

T. Iizumi et al. (eds.), *Adaptation to Climate Change in Agriculture*,
https://doi.org/10.1007/978-981-13-9235-1_14

14.1 Introduction

For any measures of agricultural adaptations to climatic change, the final goal is to get them adopted by most of the farmers or farming organizations in the target region. In the *diffusion of innovations* paradigm (hereafter, "the diffusion") (Rogers 2003), we are told that an innovation is communicated through certain channels over time among the members of a social system (Figure 1.2 of Rogers (2003)). The innovation itself – or, in this discussion, the adaptation measure – is therefore only one of the four major elements of its diffusion, with the other three being the communication channels, time, and the social system. The involvement of the other three elements explains the fact that the technical advantages of an innovation seldom warrant its successful adoption. Examples abound: the technical advantage of the Dvorak Keyboard has brought it little adoption over the QWERTY keyboard, which was designed to *slow down* typists to avoid a mechanical constraint in early typewriters (Rogers 2003). Vested interests on the side of the users and manufacturers have apparently prevailed over the technical advantage in this case. This also holds true in the technologies and practices described in the preceding chapters for the adaptation of agriculture to climate change. We need to address the question of how widely they will be adopted in addition to the question of how effective they will be when they are adopted.

The diffusion theories also tell us that many innovations originate *not* from R&D (research and development) organizations but from the users themselves (Rogers 2003). This type is called *decentralized diffusion* as opposed to *centralized diffusion* (Table 9.1 of Rogers (2003)). In centralized diffusion, a new technology developed at an R&D organization is disseminated to the users by change agents such as extension personnel. In decentralized diffusion, by contrast, the users dominate the entire process, as they serve as inventors as well as change agents and adopters (Fig. 9.1 of Rogers (2003)). In reality, most diffusions are based on a combination of the two types (Figure 9.2 of Rogers 2003).

Indeed, Fujisawa and Kobayashi (2013) reported such a case with apple growers in northern Japan. Four farmers started to grow peaches in addition to apples, and this practice was adopted by a small number of farmers for several years. This decentralized diffusion turned into a centralized one when the local government and agricultural cooperative started interventions with technical and financial supports. A much larger number of apple growers introduced the new crop, allowing them to capture a niche in the market. Because peaches are a better fit to a warmer climate than apples, the diversification of crops also made them better adapted to climatic warming in the future (Fujisawa and Kobayashi 2013).

In this chapter, we address soil frost control (SFC) as an adaptation to climate change in Tokachi, Hokkaido, Japan. SFC started as a farmer practice and diffused among the local farmers in response to climate change. It was then brought to an R&D institution, where scientists turned it into a control technology. The technology was eventually applied to 60–70% of available fields across the region through dissemination by regional farming organizations as shown in Chapter 4 of this book (Yazaki and Hirota in this volume). This is therefore another case of combined decentralized and centralized systems of diffusion in the agricultural adaptation to climate change. While technical details and the efficacy of SFC are given in Yazaki and Hirota in this volume, we focus in this chapter on the process of its development and diffusion across the two systems in the study region.

Diffusion is a dynamic process that evolves through time. We therefore describe the evolution of SFC from its initial stage through to its wide adoption, after giving the background information on the study region. We then address the question why it has been widely adopted by the farmers. The answers will be sought out in the features of the technology per se, and in the contribution of the main actors to the diffusion. Since the first author of this article, T. Hirota (TH), was heavily involved in the SFC development as one of the major actors, most descriptions of the temporal evolution and involvement of the actors are based on his direct observations and communications with the other actors.

14.2 Climate, Agriculture, and Climate Change in the Study Region

14.2.1 Climate, Agriculture, and Agricultural Organizations in Tokachi

Tokachi is a subprefectural region located on the Pacific coast of eastern Hokkaido, the northernmost island of Japan (Fig. 14.1). It has warm summers with a daily mean air temperature reaching 20°C, and cold winters with a daily minimum

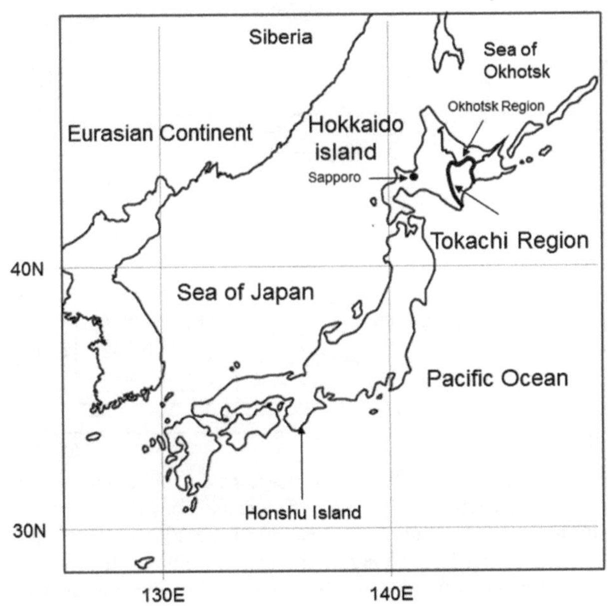

Fig. 14.1 Location of the study region: Tokachi, Hokkaido, Japan

temperature often falling below −20°C in January, the coldest month. The total annual precipitation is 800–1200 mm, of which 100–300 mm falls in winter. Snow depth in winter ranges from 0.5 to 1 m, and the topsoil is frozen from December to April.

Agriculture in Tokachi is characterized by rotation of upland crops including wheat, potatoes, sugar beets, and legumes. Farmland area per household averages over 40 ha, which is exceptionally large in comparison to the national average of 2 ha per farm household in Japan. Hokkaido accounts for nearly 80% of the total harvest of potatoes in Japan, and Tokachi produces approximately 30% of the national potato harvest.

In Tokachi, there are 24 agricultural cooperatives (JAs) with a total membership of 5843 farming households as of 2017 (Hokkaido Government Tokachi General Subprefectural Bureau 2018). The 24 JAs constitute the Tokachi Federation of Agricultural Cooperatives (TFAC) (Fig. 14.2), which serves the member JAs with testing and dissemination of new production technologies. TFAC also provides the JAs and the farmers with information services across the region.

The provincial government of Hokkaido has regional R&D institutions, including the Tokachi Agricultural Extension Center (TAEC) and the Tokachi Agricultural Experiment Station (TAES) (Fig. 14.2). The National Agriculture and Food Research Organization (NARO) conducts R&D at the Hokkaido Agricultural Research Center (HARC) (Fig. 14.2), which is headquartered in Sapporo but has its Upland Farming Division in Tokachi (Fig. 14.2). As a part of the national R&D institution, HARC is oriented more to basic research in agriculture. TAEC, by comparison, is a prefec-

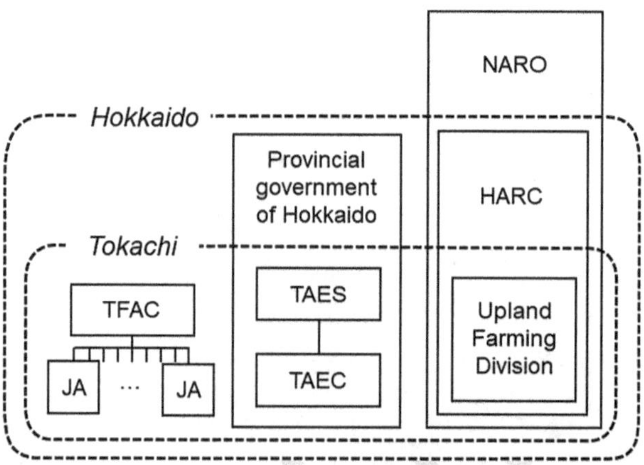

Fig. 14.2 Agricultural institutions involved in the development of the soil frost control technology in Tokachi, Hokkaido, Japan. *NARO* National Agriculture and Food Research Organization, *HARC* Hokkaido Agricultural Research Center, *TFAC* Tokachi Federation of Agricultural Cooperatives, *JA* Agricultural Cooperative, *TAES* Tokachi Agricultural Experiment Station, *TAEC* Tokachi Agricultural Extension Center. Solid lines connecting the institutions indicate close organizational links

tural R&D institution in Tokachi region, and is oriented to applications for regional agriculture.

14.2.2 Change of Climate and the Proliferation of Volunteer Potatoes

Since the late 1980s, the climate has changed in the study region. The soil surface has become covered by snow earlier and is directly exposed to cold air for a shorter period of time than before. The maximum soil frost depth has become less than 0.2 m, compared to 0.3–0.5 m through the 1980s (Hirota et al. 2006).

The shallower soil frost depth brought an unexpected and undesired effect on agriculture. Unharvested potato tubers, which had formerly been killed by soil freezing to a greater depth, can now survive the winter and emerge in the subsequent year to become a weed that hinders the growth of the next crop, e.g., legumes, sugar beets, or wheat, in rotation. Such potatoes are referred to as "volunteer potatoes" (VP) (e.g., Rahman 1980; Shirahata 2018) and must be removed to prevent competition with the crop for nutrients, to suppress pests such as potato cyst nematodes, and to prevent contamination of potato crop by the different cultivars that survived with the tubers.

Weeding VP using herbicides, rotary tillers, or manual labor was inefficient and onerous, because the VP sprout sporadically over an extended period of time. It takes farmers a long time – dozens of hours per hectare per person – to remove VP manually and repeatedly from large fields of up to several hectares during their busiest time of the farming season (Hirota et al. 2011, 2013).

14.3 Temporal Evolution of Soil Frost Control

Time is one of the major elements of the diffusion of innovations (Rogers 2003). This was also the case with SFC in Tokachi, where it took more than 30 years from the invention of the farmers' practice to the wide adoption of science-based technology by majority of farmers in the study region. Across that 30-plus years of diffusion, various actors were involved and interacted as described below and summarized in Table 14.1.

14.3.1 Progressive Farmers Developed Snowplowing Against Volunteer Potatoes

As a countermeasure to the proliferation of VP, some local farmers started snowplowing, in which they removed snow cover in the fields, exposing the soil surface to cold air (Fig. 14.3), and thereby accelerating soil freezing to kill VP (Hirota et al.

Table 14.1 Temporal evolution of the soil frost control (SFC) technology and contribution of the major actors

Period	Technology developments	Actions of major institutions and individuals[a]				Other relevant issues
		Farmers	JAs and TFAC	TAES and TAEC	HARC	
Around 1980	Snowplowing was invented	*An innovative farmer* invented snowplowing (*yukiwari*) against volunteer potatoes (VP)				
1990–2004	Reinventions were made by progressive farmers	Progressive farmers adopted *yukiwari* as VP became prevalent	Progressive JA made arrangement of *yukiwari* for farmers		**T. Hirota (TH)** developed soil temperature model (Hirota et al. 2002), and started intensive observations on atmosphere-snow-soil frost layer interactions (Hirota et al. 2005)	Late 1980s: Maximum soil frost depth has decreased due to climate change (Hirota et al. 2006)
2005	SFC became a research subject		Started to work with scientists at HARC	**A TAEC staff** brought *yukiwari* to the attention of **TH** at HARC	**TH** started collaborations with TFAC and hearing from progressive farmers	
2005–2010		Joined field experiments by JA. Progressive JA adopted SFC	Conducted field experiments. Started developing decision-support system for VP control		Identified determinants of VP proliferation (Maezuka 2008; Hirota et al. 2011). Developed soil frost control (Hirota et al. 2011)	2008: The VP problem was widely recognized by farmers and the public in Hokkaido due to news coverage of the G8 Summit

2010			Started a research project among HARC, TAES, TAEC, and TFAC.			
2010–2012		Joined the field experiments by TAES and TAEC. Disseminated SFC among farmers	Conducted on-farm demonstrations	Conducted field experiments	Applied and verified SFC in farmers' fields (Yazaki et al. 2012, 2013)	
2012–2013	SFC became available on the web	Average farmers and JAs adopted SFC	Completed SFC decision-support system	Completed SFC documentations	Completed SFC decision-support system and documentation	
2013–	SFC diffused to Okhotsk region			Transferred SFC to Okhotsk region of Hokkaido		

[a]*JA* agricultural cooperatives, *TFAC* Tokachi Federation of Agricultural Cooperatives, *TAES* Tokachi Agricultural Experiment Station, *TAEC* Tokachi Agricultural Extension Center, *HARC* Hokkaido Agricultural Research Center of the National Agricultural Research Organization. The "irreplaceable individuals" are highlighted in bold. (See text for more details of the irreplaceable individuals)

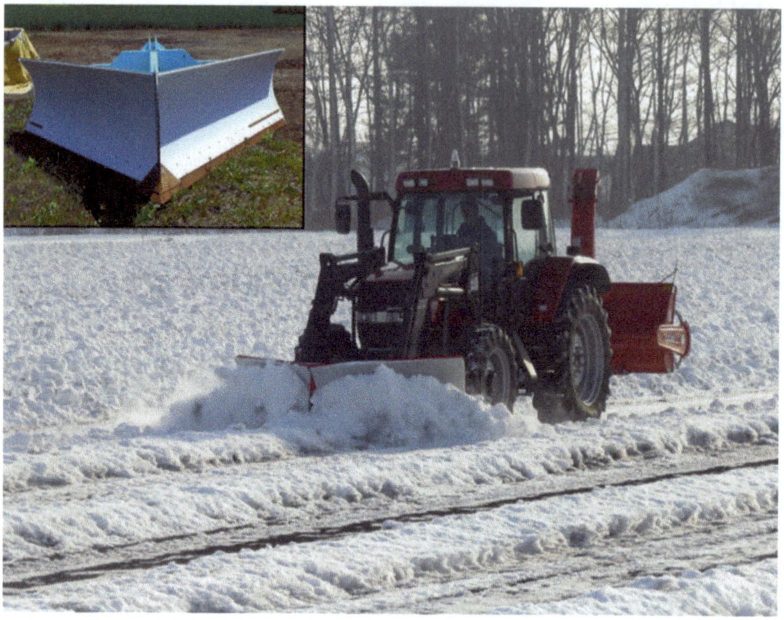

Fig. 14.3 A farmer performing *yukiwari* (snowplowing) using own tractor attached with a V-shaped blade improved for the practice (inset) in Tokachi, Hokkaido, Japan

2011). This practice was invented by a local farmer in the southern part of Tokachi during the late 1970s–1980, well before the onset of the decrease in soil frost depth. This farmer had noticed that unharvested potatoes germinated in years with deep snow cover, and conceived of the removal of snow (snowplowing) in the field to prevent VP. However, this practice was not widely adopted by other farmers at that time because VP had yet to become a major hindrance.

As the decrease of soil frost became remarkable after 1990, the snowplowing against VP gradually diffused via the network of progressive farmers. Diffusion of this practice at this early stage was initiated by the network of farmers' wives, who had the burden of manually removing VP from the fields. The farmers, on the other hand, were not strongly motivated to adopt the snowplowing idea as they were not heavily involved in the manual removal of VP but served as operators of the farm machinery.

However, possibly being prodded by their wives, they acquired it easily; they simply watched the operation by the earlier adopters and repeated it by themselves, with little need of detailed information.

As more farmers adopted snowplowing, the practice was repeatedly improved and devised by the farmers and local machinery manufacturers. Improvements were made to the hardware, i.e., snowplowing blades (Fig. 14.3 inset), and field preparations like better land-levelling beforehand. They also found that subsoiling before winter helped drainage of the snowplowed fields the next spring. It was at this time when they started to call the practice *yukiwari* in Japanese (translated into English as "snow-breaking"). It is noteworthy that the farming institutions, JAs and TAEC, did not have good grasp of the diffusion of this practice among progressive farmers at this stage. This may indicate that the institutions were on a communications network with traditional farmers that was separate from that of the progressive farmers.

Yukiwari spread among the progressive farmers in neighboring and other districts of the Tokachi region, with further improvements. At an earlier stage, the operation was limited to early winter, when the snow depth was less than 0.2 or 0.3 m, beyond which snowplowing was difficult with average farmers' tractors. However, for some farmers who owned larger machines, the deep snow was not an obstacle, and they could extend the period of snowplowing beyond early winter. Some farmers then began to contract *yukiwari* to the larger farmers.

Know-how of the technology and its application were thus accumulated among the progressive farmers through the system of decentralized diffusion (Rogers 2003) with little involvement of the agricultural institutions. Some JAs then started to make arrangements for the farmers to have access to large machines like bulldozers for *yukiwari* on contract with construction companies, which also benefitted from the new employment during the off-season of winter. *Yukiwari* thus became available for more farmers. The area of its application is estimated at several hundred hectares at the beginning of the 2000s.

14.3.2 An Extension Staff Member Brought the Farmers' Practice to the Attention of Scientists at a National R&D Institution

Yukiwari was adopted by many farmers because it effectively eliminated VP with the machine operation during the off-season of winter in place of human labor in the busy farming season. There was a basic problem with it, however. Whereas the snowplowing operation was easy for the farmers to replicate, they had no way to know the soil frost depth in their fields. They had to determine the timing and frequency of snowplowing to kill all the remaining tubers by trial and error, and relied on their intuition in response to the year-to-year fluctuation in climate. In some years, soil frost was insufficient to kill all VP, whereas in other years, too-deep soil frost delayed the start of farming in the next spring. The lack of scientific understanding about soil freezing had rendered the practice unreliable even for experienced farmers.

As previously mentioned, JAs and TAEC did not have a good grasp of *yukiwari*, and they could not give effective advice to farmers in trouble. They also did not disseminate this practice to average farmers who had not experienced it. An extension staff member at TAEC understood the basic problem of *yukiwari* and called the scientists at HARC for attention to this problem in 2005. The farmers' practice dependent on experience and intuition was thus subjected to R&D at the scientific institution.

14.3.3 Scientists Developed Soil Frost Control Technology in Collaboration with Regional and Local Farming Institutions

By the time *yukiwari* was brought to HARC, TH of HARC had developed a simple but physics-based model of soil temperature across a vertical profile (Hirota et al. 2002) and had been conducting intensive observations on atmosphere/snow layer/soil frost layer interactions (Hirota et al. 2005). TH formulated the challenge of *yukiwari* as an optimization of soil frost depth to attain two conflicting targets:

1. To attain soil freezing deep enough to kill all the remaining tubers and
2. To keep soil frost depth shallow enough to avoid excessive delay in waiting for the soil to melt the next spring

The latter target is necessary to avoid delaying the seeding of the next crops or the growth of overwintering crops in the next spring. The SFC technology was developed to optimize the soil frost depth using the soil temperature model (Hirota et al. 2011, Yazaki and Hirota in this volume). The development of SFC at HARC was facilitated by communications with various actors involving the progressive farmers, JAs, TAEC, and TAES, as noted below.

Soon after the contact with the extension staff member in 2005, TH started to communicate about *yukiwari* with progressive farmers and some JAs in Tokachi. The interviews with the progressive farmers helped him realize the diffusion of *yukiwari* and its problems as noted above. He also saw that at some JAs, they kept measuring soil frost depth and snow depth every two weeks to estimate snow melting and soil thawing for planning the start of spring farming but noted that the JAs did not have a good understanding of the diffusion of *yukiwari*.

TH was also connected to TFAC via the regional experimental station (TAES), which, as a local R&D institution, often cooperates with TFAC on dissemination of new technologies. At TFAC, they had estimated in the early 2000s that this practice had spread to several hundred hectares of fields but had no information on the details of the practice. They were therefore reluctant to disseminate it to the majority of farmers due to the lack of scientific appraisals of its efficacy, or documentation such as operation manuals for standardization of the procedure. When TH contacted them, they urged him to elaborate the practice with scientific understanding and agreed to work with him on the development of SFC technology. In cooperation with TFAC, TH and his colleagues at HARC collected the records of soil frost and snow depth observations by the JAs at 60 sites across the region and conducted a field survey of VP in 30 farmers' fields. With these observations, they established the relationship between the occurrence of VP and the soil frost depth (Maezuka 2008). They also used the observations to verify and improve the soil temperature model.

Acknowledging this progress, TFAC proposed that the scientists at HARC develop a web-based decision-making system for farmers on soil frost control. TFAC already had a web-based system for disseminating information on farming practices across the region. The scientists began field tests to develop and verify the decision-support system (Maezuka 2008). The test fields were positioned in districts where *yukiwari* had not yet been practiced, and the field tests were performed by the farmers with the leadership of the local JAs. This is a typical strategy for JA to promote adoption of new technology among the farmers.

The scientists at HARC often communicated their progress to the cooperating JAs and farmers, which facilitated the diffusion of SFC among the farmers. For example, they showed the farmers that temperature is higher in deeper soil and remaining tubers survive more therein (Fig. 3 of Hirota et al. 2011). The farmers then realized that the autumn tillage for better tilth the next spring would transfer the potatoes remaining in the surface layer into a deeper layer (Usuki 2012) and exacerbate the VP proliferation. They eventually stopped the autumn tillage in the fields after harvesting potatoes.

By coincidence, in July 2008, the 34th G8 summit was held in Hokkaido with the main theme being global warming. In association with the news reports of that political event, local TV stations chose the VP proliferation as a climate change impact and broadcast the practice of *yukiwari* widely across Hokkaido along with scientific explanations by TH. As previously mentioned, *yukiwari* is easy to understand just by watching it in operation, and the TV coverage likely accelerated its diffusion among the farmers. It is estimated that the application of *yukiwari* in the fields ranged from 1000 to 2000 ha in the period from 2008 to 2011 as estimated by TH and TAES in communications with TFAC and JAs.

14.3.4 A Regional R&D Institution Joined the Collaboration and Contributed to the Field Testing of the Soil Frost Control Technology and Its Dissemination

In 2010, the cooperation of the scientists at HARC with TFAC developed into a project funded by the Ministry of Agriculture, Forestry, and Fisheries of the Japanese national government. The regional R&D organization TAES joined this project and conducted an evaluation of the SFC technology in farmers' fields in collaboration with TAEC, the regional institution for agricultural extension. The soil temperature model was further elaborated, and the user interface in the decision support system was improved throughout the project. The major outcomes of the project were as follows:

1. The user interface of the decision support system was improved by a TFAC staff member with the farmers' point of view, rather than that of researchers, making the system easy to use for farmers and JA staff.
2. SFC with the decision support system performed better in VP control than the farmer practice based on experience and intuition (Yazaki et al. 2013).
3. Field demonstrations of SFC-based *yukiwari* conducted by TFAC and the member JAs promoted the farmers' adoption in districts where the practice had not been widely adopted.

4. The farmers involved in the field evaluation by TAEC later served as the change agents to disseminate SFC technology among neighboring farmers.
5. The experiences of the field evaluation and accumulation of scientific understanding among the TAES staff facilitated the compilation of technical manual and guidelines for SFC applications.
6. With these efforts in R&D and dissemination, the area of fields under *yukiwari* increased further to about 5000 ha in Tokachi at the end of the project in the winter of 2013, as assessed by TFAC. It is estimated that about 60–70% of farmers applied this practice to the fields after harvesting potatoes (Yazaki and Hirota in this volume).

After 2013, the technology of SFC for VP control diffused to the Okhotsk region of Hokkaido, where they have climate and farming systems similar to those in Tokachi. The transfer of the technology to Okhotsk was accompanied by technical manuals and operation guidelines via the connections among the agricultural extension centers in Hokkaido. Furthermore, the SFC turned out to be effective not only to control VP but to reduce nitrogen leaching from farmland into the groundwater during the spring melt and, at the same time, reducing the emission of a greenhouse gas, nitrous oxide (Yanai et al. 2017).

14.4 Discussion

In the evolution of SFC as noted above, we can identify the following developments that marked major steps toward the eventual adoption by the majority of farmers:

1. The snowplowing practice was first developed by farmers.
2. The farmers' practice evolved into control technology via a collaboration between scientists and regional farming organization.
3. The technology was disseminated to the farmers across the region and beyond by the farming organization and the agricultural research and extension institutions.

In what follows, we elaborate contributions of the major actors and the interactions between them to the evolution of SFC at each of the above three steps. We then draw lessons from the findings on what we could do for better adaptations to climate change.

14.4.1 SFC as a Farmer-Developed Practice

The snowplowing practice had the features of innovations that are rapidly adopted according to the diffusion theories (Rogers 2003). It is simple to understand; just watching is enough to replicate it, and it is easy to try with one's own tractor on a small scale. The result: suppression of VP is visible to everyone nearby at the next crop season. These features facilitated its adoption by farmers in addition to its efficacy in VP control.

During the initial stage of its development, the snowplowing practice was improved by progressive farmers and manufacturers in its hardware and mode of operation. As mentioned earlier, the V-shaped snowplowing blades (Fig. 14.3 inset) were found to be most effective for *yukiwari*, and better land-levelling before the snowfall was found to make the practice more effective. It is known that reinvention occurs more often in decentralized than centralized diffusions (Table 9.1 in Rogers (2003)), since the former system is controlled by an increasing number of users rather than a limited number of experts in the latter system. As an innovation diffuses in decentralized system, more users are involved in making decisions on key aspects of the diffusion including modifications to the innovation. The above improvements made at the initial stage likely contributed to the wider adoption of SFC technology at the later stage.

The critical role of the decentralized diffusion at the initial stage would be evident, if one assumes an alternative scenario in which the problem of VP was brought to the R&D institution without the earlier farmer practice. The issue would have been sent to experts in weed control, and the experts would have recommended the use of herbicides against VP as is commonly done in other potato-producing countries (Shirahata 2018). The chemical option is indeed the most straightforward solution for VP control, with wider applicability to regions outside the soil-freezing climate. With this option, however, the farmers would have missed the labor-saving aspect of SFC and had to spend extra time and money for the chemical treatment during the busy growing season. Despite the extra labor and costs, they could not have perfectly eliminated VP with herbicides, unlike SFC. Some of them might have decided not to grow potatoes to avoid VP altogether instead of spending labor for imperfect control. The other benefits of SFC, e.g., the reduction of nitrogen leaching (Yanai et al. 2017), would have also been missed with the adoption of chemical control.

The above consequences of the alternative scenario illustrate the advantage of decentralized diffusion, which is closely geared to local needs (Rogers 2003). The farmers have much better knowledge of the entirety of their own farming than anyone else, and they do not have to consider applicability to other regions that have different climates. The scientists are in a position opposite to the farmers in these respects. In adaptation to climate change with institution-led, centralized diffusion, particularly where the negative impact is already pervasive, they must find a measure within their expertise for a limited time period, rather than taking time to explore every possibility in local conditions. Because of the limited time and scope for adaptation, an institution-led adaptation tends to be reactive and progressive rather than innovative (Fujisawa et al. 2015).

There are disadvantages to decentralized diffusion, however, as compared with centralized one. One of the disadvantages is that decentralized diffusion is not appropriate for disseminating innovations that involve a high level of technical expertise (Rogers 2003). Estimation of soil frost depth under varying weather conditions was the technical expertise that was missing from the farmers' practice at the initial stage. Lack of this capability made the practice sometimes unreliable, which probably made farming organizations such as JAs reluctant to disseminate it for wider adoption.

14.4.2 Development of the Farmer Practice into the Control Technology

It was critical to the evolution of SFC technology that, in 2005, the extension staff at TAEC brought the problem of the farmers' practice to the attention of TH, who had developed the soil temperature model (Hirota et al. 2002). The model provided the means to adjust the timing and frequency of snowplowing for attaining the targeted soil frost, and thereby turned the farmers' practice based on experience and intuition into a science-based control technology that can be applied reliably. With this development, however, diffusion of SFC shifted from a decentralized system to a centralized one.

It was therefore critical that TH was also connected with TFAC via TAES. Working with TFAC, the scientists at HARC managed to cover the entire region in the VP field survey and collect observation records of soil frost. By cooperation with TFAC, the SFC technology was implemented into the web-based system that had already been developed for farmer support at TFAC. Without the existing information system, development of the decision-support system for SFC would have cost much more.

It is noteworthy that two critical links were made available to TH via his connections with the extension center (TAEC) and the experimental station (TAES). Whereas the connection with TAES was based on an institutionalized link between HARC and TAES, the connection with TAEC was interpersonal. The extension staff at TAEC had known TH's research as a member of the Hokkaido Branch of the Society of Agricultural Meteorology of Japan (SAMJ), and hence brought the subject directly to TH. The communication network offered by Hokkaido–SAMJ thus facilitated the most critical link in the evolution of the SFC technology.

14.4.3 Dissemination of the Technology to Farmers Across the Region and Beyond

As noted in the Introduction, the technical advantage of an innovation is only one of the four major elements of the diffusion. The other three elements point to the acceptance of the innovation as the critical process. In SFC, the control technology had to be disseminated to the farmers across the Tokachi region, in which TAES and TAEC played key roles along with TFAC.

The contribution of TFAC in the dissemination is evident. The demonstration of SFC in districts where it had been little adopted greatly increased the *observability* (Rogers 2003) of SFC. The practice of snowplowing is easy to replicate for those near the demonstration fields, and the web-based system at TFAC greatly reduced *complexity* in the decision-making on the practice. These traits of the technology accelerated its adoption (Rogers 2003).

TAES and TAEC contributed to dissemination of SFC by conducting technology evaluation in farmers' fields. Technical guidelines and the operation manual were established on the basis of the field evaluation and contributed to standardization of the technology, which facilitated its wide adoption by farmers. The farmers involved in the field evaluation by TAEC later served as the change agents to promote adoption of the technology by other local farmers.

The SFC technology also spread beyond the Tokachi region. This is because TAEC is a regional institution under the government of Hokkaido, and there is a network of extension centers across Hokkaido. Via this network, the SFC technology was diffused to the Okhotsk region together with the technical documents.

14.4.4 Lessons from the Development of SFC Technology

The findings in this study as well as the diffusion paradigm indicate that diffusion of innovation is a human-driven process. Among the major actors who drove the evolution of SFC technology, there were individuals who could hardly be replaced by others. The inventor of the snowplowing practice or *yukiwari* was clearly one of them. This farmer is a typical innovator, for whom the invention of VP control by *yukiwari* is only one of the various inventions and reinventions. It must be noted that he happened to invent this practice *before* the onset of climate change's impact on soil frost. Serendipity was thus involved in this invention, as with many other innovations (Rogers 2003). This farmer is also a typical *lead user* (Rogers 2003) of farm machinery, on which he makes improvements to fit his needs (Kato 2012). Studies suggest that innovations are often done by lead users instead of manufacturers in various sectors (Chapter 2 of von Hippel 2005). Therefore, the invention of *yukiwari* was not exceptional in its origin or its serendipity.

TH was another irreplaceable individual in developing the SFC technology. When he developed the model of soil temperature profile (Hirota et al. 2002), he had no intention of using it for climate change adaptation. It was only after he knew about *yukiwari* that he thought of the technological development using the model.

The two key components of the SFC technology, the snowplowing and the soil temperature model, were thus invented with no connection to one another and with no intention of them being used in climate change adaptation. They were found to be useful within the communication networks. The snowplowing first diffused through the network of progressive farmers, who adopted the practice and sometimes made reinventions. The high competency of the progressive farmers is known to facilitate the decentralized diffusion of innovation (Chapter 9 of Rogers (2003)). The soil temperature model was found to be the key component to turn *yukiwari* into a control technology after the extension staff of TAEC connected to TH. Since the link between the extension staff and TH was not institutional but interpersonal, the extension staff was also an irreplaceable individual in the technology development.

It must be noted that all the three individuals: the farmer inventor, the scientist, and the extension staff, were self-motivated rather than on duty when they took the respective action that eventually led to the development of SFC technology. Development of the SFC technology was thus closer to individual and unintended innovations threaded with serendipity than to an institution-led project with the aim of developing technologies for climate change adaptation.

The success of the SFC technology as an innovation adopted by the majority of farmers across the Tokachi region and beyond offers the following lessons:

1. We should not confine our efforts for climate change adaptation to inventions from R&D organizations to alleviate specific impacts from climate change. There may already be other innovations in farmers' networks that take advantage of their resources and that fit their needs. The R&D organizations could elaborate and/or disseminate such innovations for adoption by farmers at a larger scale.
2. We should accept the serendipity of inventions, and accordingly acknowledge that we cannot design innovations. We can, nevertheless, find values of innovations and accelerate their diffusion; a key element of this is communication networks among motivated individuals across boundaries, e.g., those between organizations, disciplines, and geographical regions.

It must be noted that the above lessons are pertinent to academic societies, whose purpose is to facilitate innovations by offering a venue of communications between the members from different organizations but sharing the same interests in scientific challenges. The Society of Agricultural Meteorology of Japan could better serve as an academic institution to facilitate innovative adaptations in agriculture to climate change by extending its communication network toward the farmers and farming organizations.

14.5 Conclusions

Based on the success of the SFC technology, we conclude the following:

1. As a farmer-originated innovation that was diffused through a decentralized system, it had traits that facilitated its rapid and sustained adoption by farmers.
2. SFC had technical advantages over other options in suppressing volunteer potatoes at high efficacy and low cost.
3. The problem of farmers' practices based on their experience and intuition was solved by turning it into a control technology based on scientific understanding.

4. Its evolution from a farmers' practice to control technology was attained by connecting the actors across disciplines and organizations. The critical connection was based on an interpersonal link via the communication network for the Hokkaido Branch of the Society of Agricultural Meteorology of Japan.
5. We should accept the serendipity of innovations and look more at their diffusion among farmers and farming organizations beyond the constrained view of institution-led technology developments.

Acknowledgments We appreciate the cooperation of the farmers in Tokachi, particularly Mr. Y. Yoshida, Mr. K. Bitou, Mr. M. Kaji, and Mr. H. Fujimori. Our appreciation also goes to Mr. M. Shirahata (formerly at TAEC), Mr. M. Suzuki, Mr. K Araki, Mr. A Iwasaki and Dr. N. Miki of the Hokkaido Research Organization (TAES), and Mr. K. Maezuka, Mr. T. Kaji, Mr. A. Sawasaki, and Mr. T. Taraba of TFAC, for their cooperation. This research was supported by the Project of the NARO Bio-oriented Technology Research Advancement Institution (research program on development of innovative technology [29017C]) and J.S.P.S KAKENHI Grant number 19H00963.

References

Fujisawa M, Kobayashi K (2013) Shifting from apple to peach farming in Kazuno, northern Japan: perceptions of and responses to climatic and non-climatic impacts. Reg Environ Chang 13:1211–1222

Fujisawa M, Kobayashi K, Johnston P, New M (2015) What drives farmers to make top-down or bottom-up adaptation to climate change and fluctuations? A comparative study on three cases of apple farming in Japan and South Africa. PLoS One 10(3):e0120563. https://doi.org/10.1371/journal.pone.0120563

Hirota T, Pomeroy JW, Granger RJ, Maule CP (2002) An extension of the force-restore method to estimating soil temperature at depth and evaluation for frozen soils under snow. J Geophys Res 107:4767. https://doi.org/10.1029/2001JD001280

Hirota T, Iwata Y, Hamasaki T, Sameshima R, Hayashi M (2005) Micrometeorological conditions and the thermal and moisture characteristics of seasonally frozen soil in Eastern Hokkaido: new comprehensive hydro-meteorological observation system for atmosphere-snow-frozen soil interaction. J Agric Meteorol 60:673–676

Hirota T, Iwata Y, Hayashi M, Suzuki S, Hamasaki T, Sameshima R, Takayabu I (2006) Decreasing soil-frost depth and its relation to climate change in Tokachi, Hokkaido, Japan. J Meteorol Soc Jpn 84:821–833

Hirota T, Usuki K, Hayashi M, Nemoto M, Iwata Y, Yanai Y, Yazaki T, Inoue S (2011) Soil frost control: agricultural adaptation to climate variability in a cold region of Japan. Mitig Adapt Strateg Glob Change 16:791–802

Hirota T, Yazaki T, Usuki K, Hayashi M, Nemoto M, Iwata Y, Yanai Y, Inoue S, Suzuki T, Shirahata M, Kajiyama T, Araki K, Maezuka K (2013) Soil frost control: its application to volunteer potato management in a cold region. In: Imai R, Yoshida M, Matsumoto N (eds) Plant and microbe adaptations to cold in a changing world. Springer, New York, pp 51–61

Hokkaido Government Tokachi General Subprefectural Bureau (2018) 2017 Agriculture in Tokachi. http://www.tokachi.pref.hokkaido.lg.jp/ss/num/2017_tokachi.htm/. Accessed 12 Jan 2019

Kato Y (2012) Original machineries supporting the family farming (translation). Nogyo Keieisha, 2012 July:62–63 (in Japanese)

Maezuka K (2008) The volunteer potatoes problems and its managements by snow removal. J Agric Meteorol Hokkaido 60:39–44. (in Japanese)

Rahman A (1980) Biology and control of volunteer potatoes – a review. NZ J Exp Agric 8:313–319
Rogers EM (2003) Diffusion of innovations, 5th edn. Free Press, New York
Shirahata M (2018) The review about volunteer potato. J Agric Meteorol Hokkaido 70:12–21. (in Japanese)
Usuki K (2012) Control of volunteer potato occurrence using freezing behavior of tubers: relations to climate resources. Hokuno 79:180–185. (in Japanese)
Von Hippel E (2005) Democratizing innovation. The MIT Press, Cambridge
Yanai Y, Iwata Y, Hirota T (2017) Optimum soil frost depth to alleviate climate change effects in cold region agriculture. Sci Rep 7:44860. https://doi.org/10.1038/srep44860
Yazaki T, Hirota T (this volume) Controlling the depth of soil frost in farm fields in Japan. In: Iizumi T, Hirata R, Matsuda R (eds) Adaptation to climate change in agriculture. Springer Nature, Singapore
Yazaki T, Hirota T, Iwata Y, Inoue S, Usuki K, Suzuki T, Shirahata M, Iwasaki A, Kajiyama T, Araki K, Takamiya Y, Maezuka K (2013) Effective killing of volunteer potato (*Solanum tuberosum L.*) tubers by soil frost control using agrometeorological information: an adaptive countermeasure to climate change in a cold region. Agric For Meteorol 182–183:91–100

Kolibri 360 GmbH
Ferdinand-Jühlke-Straße 7
99095 Erfurt, Deutschland
produktsicherheit@kolibri360.de